Distributed Data Systems with Azure Databricks

Create, deploy, and manage enterprise
data pipelines

Alan Bernardo Palacio

BIRMINGHAM—MUMBAI

Distributed Data Systems with Azure Databricks

Copyright © 2021 Packt Publishing

Group Product Manager: Kunal Parikh

Publishing Product Manager: Ali Abidi

Senior Editor: David Sugarman

Content Development Editor: Joseph Sunil

Technical Editor: Manikandan Kurup

Copy Editor: Safis Editing

Project Coordinator: Aparna Nair

Proofreader: Safis Editing

Indexer: Rekha Nair

Production Designer: Vijay Kamble

First published: May 2021

Production reference: 1210521

Published by Packt Publishing Ltd.
Livery Place
35 Livery Street
Birmingham
B3 2PB, UK.

ISBN 978-1-83864-721-6

www.packt.com

Contributors

About the author

Alan Bernardo Palacio is a senior data engineer at Ernst and Young. He is an accomplished data scientist with a master's degree in modeling for science and engineering, and he is also a mechanical engineer. He has worked on various projects, such as machine translation projects for Disney and computer vision and natural language processing models at Ernst and Young.

About the reviewer

Adwait Ullal is a technology consultant based in Silicon Valley. He works with Fortune 500 companies to provide cloud and enterprise architecture guidance. Adwait's prior experience includes application and solutions architecture, specializing in Microsoft technologies. Adwait has presented on cloud and enterprise architecture topics at local code camps and meetups.

Table of Contents

2

Creating an Azure Databricks Workspace

Section 2: Data Pipelines with Databricks

3

Creating ETL Operations with Azure Databricks

4

Delta Lake with Azure Databricks

5

Introducing Delta Engine

6

Introducing Structured Streaming

Section 3: Machine and Deep Learning with Databricks

7
Using Python Libraries in Azure Databricks

8
Databricks Runtime for Machine Learning

9

Databricks Runtime for Deep Learning

10

Model Tracking and Tuning in Azure Databricks

11

Managing and Serving Models with MLflow and MLeap

12

Distributed Deep Learning in Azure Databricks

Other Books You May Enjoy

Index

Preface

Microsoft Azure Databricks helps you to harness the power of distributed computing and apply it to create robust data pipelines, along with training and deploying **Machine Learning (ML)** and **Deep Learning (DL)** models. Databricks' advanced features enable developers to process, transform, and explore data. Distributed data systems with Azure Databricks will help you to put your knowledge of Databricks to work to create big data pipelines.

The book provides a hands-on approach to implementing Azure Databricks and its associated methodologies that will make you productive in no time. Complete with detailed explanations of essential concepts, practical examples, and self-assessment questions, you'll begin with a quick introduction to Databricks' core functionalities, before performing distributed model training and inference using TensorFlow and Spark MLlib. As you advance, you'll explore MLflow Model Serving on Azure Databricks and distributed training pipelines using HorovodRunner in Databricks. Finally, you'll discover how to transform, use, and obtain insights from massive amounts of data to train predictive models and create entire fully working data pipelines.

By the end of this MS Azure book, you'll have gained a solid understanding of how to work with Databricks to create and manage an entire big data pipeline.

Who this book is for

This book is for software engineers, ML engineers, data scientists, and data engineers who are new to Azure Databricks and who want to build high-quality data pipelines without worrying about infrastructure. Knowledge of the basics of Azure Databricks is required to learn the concepts covered in this book more effectively. A basic understanding of ML concepts and beginner-level Python programming knowledge is also recommended.

What this book covers

Chapter 1, Introduction to Azure Databricks, takes you through the core functionalities of Databricks, including how we interact with the workspace environment, a quick look into the main applications, and how we will be using the tool for Python users. This covers topics such as workspace, interface, computation management, and Databricks notebooks.

Chapter 2, Creating an Azure Databricks Workspace, teaches you how to apply all the previous concepts using the different tools that Azure has in order to interact with the workspace. This includes using PowerShell and the Azure CLI to manage all Databricks' resources.

Chapter 3, Creating ETL Operations with Azure Databricks, shows you how to manage different data sources, transform them, and create an entire event-driven ETL.

Chapter 4, Delta Lake with Azure Databricks, explores Delta Lake and how to implement it for various operations.

Chapter 5, Introducing Delta Engine, explores Delta Engine and also shows you how to use it along with Delta Lake and create efficient ETLs in Databricks.

Chapter 6, Introducing Structured Streaming, provides explanations on notebooks, details on how to use specific types of streaming sources and sinks, how to put streaming into production, and notebooks demonstrating example use cases.

Chapter 7, Using Python Libraries in Azure Databricks, explores all the nuances regarding working with Python, as well as introducing core concepts regarding models and data that will be studied in more detail later on.

Chapter 8, Databricks Runtime for Machine Learning, acts as a deep dive for us in the development of classic ML algorithms to train and deploy models based on tabular data, all while exploring libraries and algorithms as well. The examples will be focused on the particularities and advantages of using Databricks for ML.

Chapter 9, Databricks Runtime for Deep Learning, acts as a deep dive for us in the development of classic DL algorithms to train and deploy models based on unstructured data, all while exploring libraries and algorithms as well. The examples will be focused on the particularities and advantages of using Databricks for DL.

Chapter 10, Model Tracking and Tuning in Azure Databricks, focuses on model tuning, deployment, and control using Databricks' functionalities, such as AutoML and Delta Lake, while using it in conjunction with popular libraries such as TensorFlow.

Chapter 11, Managing and Serving Models with MLflow and MLeap, explores in more detail the MLflow library, an open source platform for managing the end-to-end ML life cycle. This library allows the user to track experiments, record and compare parameters, centralize model storage, and more. You will learn how to use it in combination with what was learned in the previous chapters.

Chapter 12, Distributed Deep Learning in Azure Databricks, demonstrates how to use Horovod to make distributed DL faster by taking single-GPU training scripts and scaling them to train across many GPUs in parallel.

To get the most out of this book

You need a functional PC with an internet connection and an Azure Services subscription.

Download the example code files

You can download the example code files for this book from GitHub at `https://github.com/PacktPublishing/Distributed-Data-Systems-with-Azure-Databricks`. In case there's an update to the code, it will be updated on the existing GitHub repository.

We also have other code bundles from our rich catalog of books and videos available at `https://github.com/PacktPublishing/`. Check them out!

Download the color images

We also provide a PDF file that has color images of the screenshots/diagrams used in this book. You can download it here: `http://www.packtpub.com/sites/default/files/downloads/9781838647216_ColorImages.pdf`..

Conventions used

There are a number of text conventions used throughout this book.

`Code in text`: Indicates code words in text, database table names, folder names, filenames, file extensions, pathnames, dummy URLs, user input, and Twitter handles. Here is an example: "When using Azure Databricks Runtime ML, we have the option to use the `dbfs:/ml` folder."

A block of code is set as follows:

```
diamonds = spark.read.format('csv').options(header='true',
inferSchema='true').load('/databricks-datasets/Rdatasets/data-
001/csv/ggplot2/diamonds.csv')
```

> **Tips or important notes**
> Appear like this.

Get in touch

Feedback from our readers is always welcome.

General feedback: If you have questions about any aspect of this book, mention the book title in the subject of your message and email us at customercare@packtpub.com.

Errata: Although we have taken every care to ensure the accuracy of our content, mistakes do happen. If you have found a mistake in this book, we would be grateful if you would report this to us. Please visit www.packtpub.com/support/errata, selecting your book, clicking on the Errata Submission Form link, and entering the details.

Piracy: If you come across any illegal copies of our works in any form on the internet, we would be grateful if you would provide us with the location address or website name. Please contact us at copyright@packt.com with a link to the material.

If you are interested in becoming an author: If there is a topic that you have expertise in, and you are interested in either writing or contributing to a book, please visit authors. packtpub.com.

Reviews

Please leave a review. Once you have read and used this book, why not leave a review on the site that you purchased it from? Potential readers can then see and use your unbiased opinion to make purchase decisions, we at Packt can understand what you think about our products, and our authors can see your feedback on their book. Thank you!

For more information about Packt, please visit packt.com.

Section 1: Introducing Databricks

This section introduces Databricks for new users and discusses its functionalities as well as the advantages that we have while dealing with massive amounts of data.

This section contains the following chapters:

1
Introduction to Azure Databricks

Modern information systems work with massive amounts of data, with a constant flow that increases every day at an exponential rate. This flow comes from different sources, including sales information, transactional data, social media, and more. Organizations have to work with this information in processes that include transformation and aggregation to develop applications that seek to extract value from this data.

Apache Spark was developed to process this massive amount of data. Azure Databricks is built on top of Apache Spark, abstracting most of the complexities of implementing it, and with all the benefits that come with integration with other Azure services. This book aims to provide an introduction to Azure Databricks and explore the applications it has in modern data pipelines to transform, visualize, and extract insights from large amounts of data in a distributed computation environment.

In this introductory chapter, we will explore these topics:

- Introducing Apache Spark
- Introducing Azure Databricks
- Discovering core concepts and terminology
- Interacting with the Azure Databricks workspace
- Using Azure Databricks notebooks

- Exploring data management

- Exploring computation management

- Exploring authentication and authorization

These concepts will help us to later understand all of the aspects of the execution of our jobs in Azure Databricks and to move easily between all its assets.

Technical requirements

To understand the topics presented in this book, you must be familiar with data science and data engineering terms, and have a good understanding of Python, which is the main programming language used in this book, although we will also use SQL to make queries on views and tables.

In terms of the resources required, to execute the steps in this section and those presented in this book, you will require an Azure account as well as an active subscription. Bear in mind that this is a service that is paid, so you will have to introduce your credit card details to create an account. When you create a new account, you will receive a certain amount of free credit, but there are certain options that are limited to premium users. Always remember to stop all the services if you are not using them.

Introducing Apache Spark

To work with the huge amount of information available to modern consumers, **Apache Spark** was created. It is a distributed, cluster-based computing system and a highly popular framework used for big data, with capabilities that provide speed and ease of use, and includes APIs that support the following use cases:

- Easy cluster management

- Data integration and ETL procedures

- Interactive advanced analytics

- ML and deep learning

- Real-time data processing

It can run very quickly on large datasets thanks to its in-memory processing design that allows it to run with very few read/write disk operations. It has a SQL-like interface and its object-oriented design makes it very easy to understand and write code for; it also has a large support community.

Despite its numerous benefits, Apache Spark has its limitations. These limitations include the following:

- Users need to provide a database infrastructure to store the information to work with.

- The in-memory processing feature makes it fast to run, but also implies that it has high memory requirements.

- It isn't well suited for real-time analytics.

- It has an inherent complexity with a significant learning curve.

- Because of its open source nature, it lacks dedicated training and customer support.

Let's look at the solution to these issues: Azure Databricks.

Introducing Azure Databricks

With these and other limitations in mind, **Databricks** was designed. It is a cloud-based platform that uses Apache Spark as a backend and builds on top of it, to add features including the following:

- Highly reliable data pipelines

- Data science at scale

- Simple data lake integration

- Built-in security

- Automatic cluster management

Built as a joint effort by Microsoft and the team that started Apache Spark, Azure Databricks also allows easy integration with other Azure products, such as **Blob Storage** and **SQL databases**, alongside **AWS services**, including **S3 buckets**. It has a dedicated support team that assists the platform's clients.

Databricks streamlines and simplifies the setup and maintenance of clusters while supporting different languages, such as **Scala** and **Python**, making it easy for developers to create ETL pipelines. It also allows data teams to have real-time, cross-functional collaboration thanks to its notebook-like integrated workspace, while keeping a significant amount of backend services managed by Azure Databricks. Notebooks can be used to create jobs that can later be scheduled, meaning that locally developed notebooks can be deployed to production easily. Other features that make Azure Databricks a great tool for any data team include the following:

- A high-speed connection to all Azure resources, such as storage accounts.
- Clusters scale and are terminated automatically according to use.
- The optimization of SQL.
- Integration with BI tools such as Power BI and Tableau.

Let's examine the architecture of Databricks next.

Examining the architecture of Databricks

Each Databricks cluster is a Databricks application composed of a set of pre-configured, VMs running as Azure resources managed as a single group. You can specify the number and type of VMs that it will use while Databricks manages other parameters in the backend. The managed resource group is deployed and populated with a virtual network called **VNet**, a security group that manages the permissions of the resources, and a storage account that will be used, among other things, as the Databricks filesystem. Once everything is deployed, users can manage these clusters through the Azure Databricks UI. All the metadata used is stored in a geo-replicated and fault-tolerant Azure database. This can all be seen in *Figure 1.1*:

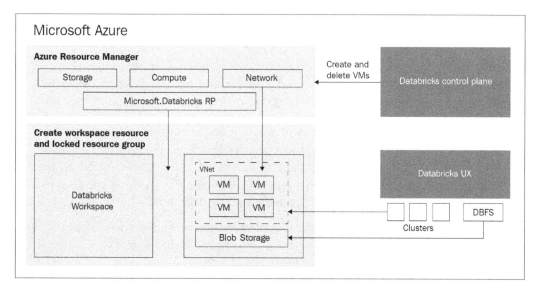

Figure 1.1 – Databricks architecture

The immediate benefit this architecture gives to users is that there is a seamless connection with Azure, allowing them to easily connect Azure Databricks to any resource within the same Azure account and have a centrally managed Databricks from the Azure control center with no additional setup.

As mentioned previously, Azure Databricks is a managed application on the Azure cloud that is composed by a **control plane** and a **data plane**. The control plane is on the Azure cloud and hosts services such as cluster management and jobs services. The data plane is a component that includes the aforementioned VNet, NSG, and the storage account that is known as DBFS.

You could also deploy the data plane in a customer-managed VNet to allow data engineering teams to build and secure the network architecture according to their organization policies. This is called VNet injection.

Now that we have seen how everything is laid out under the hood, let's discuss some of the core concepts behind Databricks.

Discovering core concepts and terminology

Before diving into the specifics of how to create our cluster and start working with Databricks, there are a certain number of concepts with which we must familiarize ourselves first. Together, these define the fundamental tools that Databricks provides to the user and are available both in the web application UI as well as the REST API:

- **Workspaces**: An Azure Databricks workspace is an environment where the user can access all of their assets: jobs, notebooks, clusters, libraries, data, and models. Everything is organized into folders and this allows the user to save notebooks and libraries and share them with other users to collaborate. The workspace is used to store notebooks and libraries, but not to connect or store data.

- **Data**: Data can be imported into the mounted Azure Databricks distributed filesystem from a variety of sources. This can be uploaded as tables directly into the workspace, from Azure Blob Storage or AWS S3.

- **Notebooks**: Databricks notebooks are very similar to Jupyter notebooks in Python. They are web interface applications that are designed to run code thanks to runnable cells that operate on files and tables, and that also provide visualizations and contain narrative text. The end result is a document with code, visualizations, and clear text documentation that can be easily shared. Notebooks are one of the two ways that we can run code in Azure Databricks. The other way is through jobs. Notebooks have a set of cells that allow the user to execute commands and can hold code in languages such as Scala, Python, R, SQL, or Markdown. To be able to execute commands, they have to be connected to a cluster, but this connection is not necessarily permanent. This allows an easy way to share these notebooks via the web or in a local machine. Notebooks can be scheduled and triggered as jobs to create a data pipeline, run ML models, or update dashboards:

≡   Q   5 BroadcastHashJoin - py (Python)   ⊙   ?   ▲

BroadcastHashJoin

- Configuring a BroadcastHashJoin is a way to optimize joining a large and a small table in Spark SQL.
- This notebook will cover the how to configure a BroadcastHashJoin and why to choose it over a ShuffledHashJoin.

Setup: Create a large table that will be joined with a smaller table.

```python
from pyspark.sql import Row

array = []
for i in range(0, 1000000):
    array.append(Row(num=i, bit = i % 2))

dataFrame = sqlContext.createDataFrame(sc.parallelize(array))
dataFrame.repartition(100).registerTempTable("my_large_table")
```

Command took 7.72s

```python
display(array)
```

Figure 1.2 – Azure Databricks notebook. Source: `https://databricks.com/wp-content/uploads/2015/10/notebook-example.png`

- **Clusters**: A cluster is a set of connected servers that work together collaboratively as if they are a single (much more powerful) computer. In this environment, you can perform tasks and execute code from notebooks working with data stored in a certain storage facility or uploaded as a table. These clusters have the means to manage and control who can access each one of them. Clusters are used to improve performance and availability compared to a single server, while typically being more cost-effective than a single server of comparable speed or availability. It is in the clusters where we run our data science jobs, ETL pipelines, analytics, and more.

There is a distinction between all-purpose clusters and job clusters. **All-purpose clusters** are where we work collaboratively and interactively using notebooks, but **job clusters** are where we execute automatic and more concrete jobs. The way of creating these clusters differs depending on whether it is an all-purpose cluster or a job cluster. The former can be created using the UI, CLI, or REST API, while the latter is created using the job scheduler to run a specific job and is terminated when this is done.

- **Jobs**: Jobs are the tasks that we run when executing a notebook, JAR, or Python file in a certain cluster. The execution can be created and scheduled manually or by the REST API.

- **Apps**: Third-party apps such as Table can be used inside Azure Databricks. These integrations are called apps.

- **Apache SparkContext/environments**: Apache SparkContext is the main application in Apache Spark running internal services and connecting to the Spark execution environment. While, historically, Apache Spark has had two core contexts available to the user (SparkContext and SQLContext), in the 2.X versions, there is just one – the **SparkSession**.

- **Dashboards**: Dashboards are a way to display the output of the cells of a notebook without the code that is required to generate them. They can be created from notebooks:

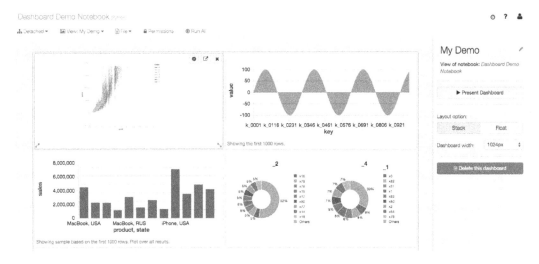

Figure 1.3 – Azure Databricks notebook. Source: `https://databricks.com/wp-content/uploads/2016/02/Databricks-dashboards-screenshot.png`

- **Libraries**: Libraries are modules that add functionality, written in Scala or Python, that can be pulled from a repository or installed via package management systems utilities such as PyPI or Maven.

- **Tables**: Tables are structured data that you can use for analysis or for building models that can be stored on Amazon S3 or Azure Blob Storage, or in the cluster that you're currently using cached in memory. These tables can be either global or local, the first being available across all clusters. A local table cannot be accessed from other clusters.

- **Experiments**: Every time we run MLflow, it belongs to a certain experiment. Experiments are the central way of organizing and controlling all the MLflow runs. In each experiment, the user can search, compare, and visualize results, as well as downloading artifacts or metadata for further analysis.

- **Models**: While working with ML or deep learning, the models that we train and use to infer are registered in the Azure Databricks MLflow Model Registry. MLflow is an open source platform designed to manage ML life cycles, which includes the tracking of experiments and runs, and MLflow Model Registry is a centralized model store that allows users to fully control the life cycle of MLflow models. It has features that enable us to manage versions, transition between different stages, have a chronological model heritage, and control model version annotations and descriptions.

- **Azure Databricks workspace filesystem**: Azure Databricks is deployed with a distributed filesystem. This system is mounted in the workspace and allows the user to mount storage objects and interact with them using filesystem paths. It allows us to persist files so the data is not lost when the cluster is terminated.

This section focused on the core pieces of Azure Databricks. In the next section, you will learn how to interact with Azure Databricks through the workspace, which is the place where we interact with our assets.

Interacting with the Azure Databricks workspace

The Azure Databricks **workspace** is where you can manage objects such as notebooks, libraries, and experiments. It is organized into folders and it also provides access to data, clusters, and jobs:

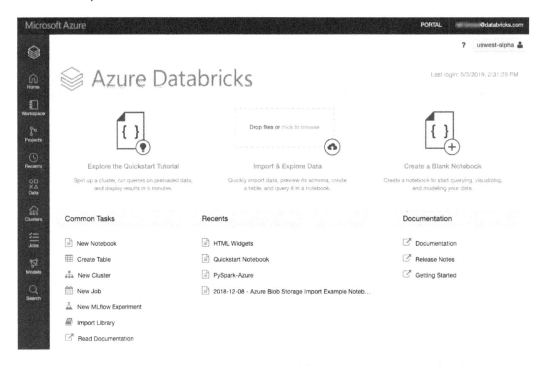

Figure 1.4 – Databricks workspace. Source: `https://docs.microsoft.com/en-us/azure/databricks/workspace/`

Access and control of a workspace and its assets can be made through the UI, CLI, or API. We will focus on using the UI.

Workspace assets

In the Azure Databricks workspace, you can manage different assets, most of which we have discussed in the terminology. These assets are as follows:

- Clusters
- Notebooks
- Jobs

- Libraries
- Assets folders
- Models
- Experiments

In the following sections, we will dive deeper into how to work with folders and other workspaces objects. The management of these objects is central to running our tasks in Azure Databricks.

Folders

All of our static assets within a workspace are stored in folders within the workspace. The stored assets can be notebooks, libraries, experiments, and other folders. Different icons are used to represent folders, notebooks, directories, or experiments. Click a directory to deploy the drop-down list of items:

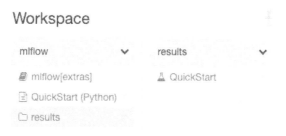

Figure 1.5 – Workspace folders

Clicking on the drop-down arrow in the top-right corner will unfold the menu item, allowing the user to perform actions with that specific folder:

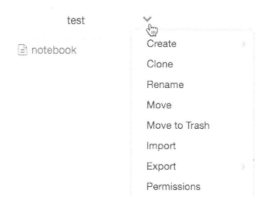

Figure 1.6 – Workspace folders drop-down menu

Special folders

The Azure Databricks workspace has three special folders that you cannot rename or move to a special folder. These special folders are as follows:

- Workspace
- Shared
- Users

Workspace root folder

The **Workspace** root folder is a folder that contains all of your static assets. To navigate to this folder, click the workspace or home icon and then click the go back icon:

Figure 1.7 – Workspace root folder

Within the **Workspace** root folder, you either select **Shared** or **Users**. The former is for sharing objects with other users that belong to your organization, and the latter contains a folder for a specific user.

By default, the **Workspace** root folder and all of its contents are available for all users, but you can control and manage access by enabling workspace access control and setting permissions.

User home folders

Within your organization, every user has their own directory, which will be their root directory:

Figure 1.8 – Workspace Users folder

Objects in a user folder will be private to a specific user if workspace access control is enabled. If a user's permissions are removed, they will still be able to access their home folder.

Workspace object operations

To perform an action on a workspace object, right-click the object or click the drop-down icon at the right side of an object to deploy the drop-down menu:

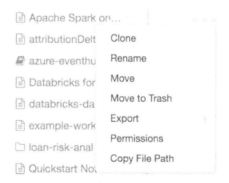

Figure 1.9 – Operations on objects in the workspace

If the object is a folder, from this menu, the user can do the following:

- Create a notebook, library, MLflow experiment, or folder.
- Import a Databricks archive.

If it is an object, the user can choose to do the following:

- Clone the object.

- Rename the object.

- Move the object to another folder.

- Move the object to Trash.

- Export a folder or notebook as a Databricks archive.

- If the object is a notebook, copy the notebook's file path.

- If you have Workspace access control enabled, set permissions on the object.

When the user deletes an object, this object goes to the Trash folder, in which everything is deleted after 30 days. Objects can be restored from the Trash folder or be eliminated permanently.

Now that you have learned how to interact with Azure Databricks assets, we can start working with Azure Databricks notebooks to manipulate data, create ETLs, ML experiments, and more.

Using Azure Databricks notebooks

In this section, we will describe the basics of working with notebooks within Azure Databricks.

Creating and managing notebooks

There are different ways to interact with notebooks in Azure Databricks. We can either access them through the UI using CLI commands, or by means of the workspace API. We will focus on the UI for now:

1. By clicking on the **Workspace** or **Home** button in the sidebar, select the drop-down icon next to the folder in which we will create the notebook. In the **Create Notebook** dialog, we will choose a name for the notebook and select the default language:

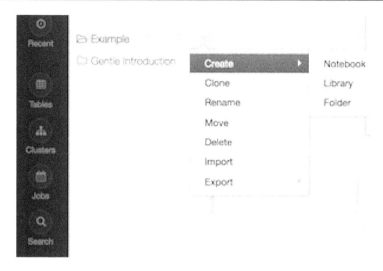

Figure 1.10 – Creating a new notebook

2. Running clusters will show notebooks attached to them. We can select one of them to attach the new notebook to; otherwise, we can attach it once the notebook has been created in a specific location.

3. To open a notebook, in your workspace, click on the icon corresponding to the notebook you want to open. The notebook path will be displayed when you hover over the notebook title.

> **Note**
> If you have an Azure Databricks Premium plan, you can apply access control to the workspace assets.

External notebook formats

Azure Databricks supports several notebook formats, which can be scripts in one of the supported languages (Python, Scala, SQL, and R), HTML documents, DBC archives (Databricks native file format), IPYNB Jupyter notebooks, and R Markdown documents.

Importing a notebook

We can import notebooks into the Azure workspace by clicking in the drop-down menu and selecting **Import**. After this, we can specify either a file or a URL that contains the file in one of the supported formats and then click **Import**:

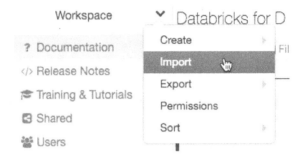

Figure 1.11 – Importing a notebook into the workspace

Exporting a notebook

You can export a notebook in one of the supported file formats by clicking on the **File** button in the notebook toolbar and then selecting **Export**. Bear in mind that the results of each cell will be included if you have not cleared them.

Notebooks and clusters

To be able to work, a notebook needs to be attached to a running cluster. We will now learn about how notebooks connect to the clusters and how to manage these executions.

Execution contexts

When a notebook is attached to a cluster, a **read-eval-print-loop** (**REPL**) environment is created. This environment is specific to each one of the supported languages and is contained in an execution context.

There is a limit of 145 execution contexts running in a single cluster. Once that number is reached, you cannot attach any more notebooks to that cluster or create a new execution context.

Idle execution contexts

If an execution context has passed a certain time threshold without any executions, it is considered idle and automatically detached from the notebook. This threshold is, by default, 25 hours.

One thing to consider is that when a cluster reaches its maximum context limit, Azure Databricks will remove the least recently used idle execution contexts. This is called an **eviction**.

If a notebook gets evicted from the cluster it was attached to, the UI will display a message:

Figure 1.12 – Detached notebook notification

We can configure this behavior when creating the cluster or we can disable it by setting the following:

```
spark.databricks.chauffeur.enableIdleContextTracking false
```

Attaching a notebook to a cluster

Notebooks are attached to a cluster by selecting one from the drop-down menu in the notebook toolbar.

A notebook attached to a running cluster has the following Spark environment variables by default:

Class	Variable Name
SparkContext	sc
SQLContext/HiveContext	sqlContext
SparkSession (Spark 2.x)	spark

Figure 1.13 – A table showing Spark environment variables

We can check the Spark version running in the cluster where the notebook is attached by running the following Python code in one of the cells:

```
spark.version
```

We can also see the current Databricks runtime version with the following command:

```
spark.conf.get("spark.databricks.clusterUsageTags.
sparkVersion")
```

These properties are required by the Clusters and Jobs APIs to communicate between themselves.

On the cluster details page, the **Notebooks** tab will show all the notebooks attached to the cluster, as well as the status and the last time it was used:

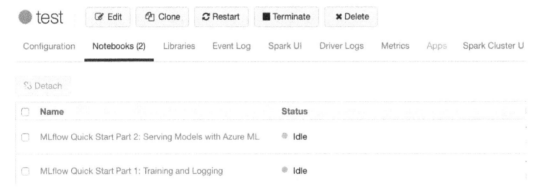

Figure 1.14 – Notebooks attached to a cluster

Attaching a notebook to a cluster is necessary in order to make them work; otherwise, we won't be able to execute the code in it.

Notebooks are detached from a cluster by clicking in the currently attached cluster and selecting **Detach**:

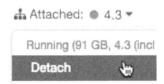

Figure 1.15 – Detaching a notebook from a cluster

This causes the cluster to lose all the values stored as variables in that notebook. It is good practice to always detach the notebooks from the cluster once we have finished working on them. This prevents the autostopping of running clusters, in case there is a process running in the notebook (which could cause undesired costs).

Scheduling a notebook

As mentioned before, notebooks can be scheduled to be executed periodically. To schedule a notebook job to run periodically, click the **Schedule** button at the top right of the notebook toolbar.

A notebook's core functionalities

Now, we'll look at how you can use a notebook.

Notebook toolbar

Notebooks have a toolbar that contains information on the cluster to which it is attached, and to perform actions such as exporting the notebook or changing the predefined language (depending on the Databricks runtime version):

Figure 1.16 – Notebook toolbar

This toolbar helps us to navigate the general options in our notebook and makes it easier to manage how we interact with the computation cluster.

Cells

Cells have code that can be executed:

Figure 1.17 – Execution cells

At the top-left corner of a cell, in the cell actions, you have the following options: **Run this cell**, **Dashboard**, **Edit**, **Hide,** and **Delete**:

- You can use the **Undo** keyboard shortcut to restore a deleted cell by selecting **Undo Delete Cell** from **Edit**.

- Cells can be cut using cell actions or the **Cut** keyboard shortcut.

- Cells are added by clicking on the **Plus** icon at the bottom of each cell or by selecting **Add Cell Above** or **Add Cell Below** from the cell menu in the notebook toolbar.

Running cells

Specific cells can be run from the cell actions toolbar. To run several cells, we can choose between **Run all**, **all above**, or **all below**. We can also select **Run All**, **Run All Above**, or **Run All Below** from the **Run** option in the notebook toolbar. Bear in mind that **Run All Below** includes the cells you are currently in.

Default language

The default language for each notebook is shown in parentheses next to the notebook name, which, in the following example, is SQL:

Figure 1.18 – Cell default language

If you click the name of the language in parentheses, you will be prompted by a dialog box in which you can change the default language of the notebook:

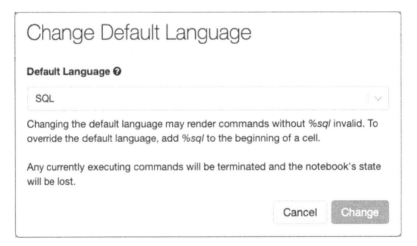

Figure 1.19 – Changing the default language of a cell

When the default language is changed, magic commands will be added to the cells that are not in the new default language in order to keep them working.

The language can also be specified in each cell by using the magic commands. Four magic commands are supported for language specification: %python, %r, %scala, and %sql.

There are also other magic commands such as %sh, which allows you to run shell code; %fs to use dbutils filesystem commands; and %md to specify Markdown, for including comments and documentation. We will look at this in a bit more detail.

Including documentation

Markdown is a lightweight markup language with plain text-formatting syntax, often used for formatting readme files, which allows the creation of rich text using plain text.

As we have seen before, Azure Databricks allows Markdown to be used for documentation by using the %md magic command. The markup is then rendered into HTML with the desired formatting. For example, the next code is used to format text as a title:

```
%md # Hello This is a Title
```

It is rendered as an HTML title:

Figure 1.20 – Markdown title

Documentation blocks are one of the most important features of Azure Databricks notebooks. They allow us to state the purpose of our code and how we interpret our results.

Command comments

Users can add comments to specific portions of code by highlighting it and clicking on the comment button in the bottom-right corner of the cell:

Figure 1.21 – Selecting a portion of code

This will prompt a textbox in which we can place comments to be reviewed by other users. Afterward, the commented text will be highlighted:

Figure 1.22 – Commenting on the selection

Comments allow us to propose changes or require information on specific portions of the notebook without intervening in the content.

Downloading a cell result

You can download the tabular results from a cell to your local machine by clicking on the download button at the bottom of a cell:

Figure 1.23 – Downloading full results from a cell

By default, Azure Databricks limits you to viewing 1,000 rows of a DataFrame, but if there is more data present, we can click on the drop-down icon and select **Download full results** to see more.

Formatting SQL

Formatting SQL code can take up a lot of time, and enforcing standards across notebooks can be difficult.

Azure Databricks has a functionality for formatting SQL code in notebook cells, so as to reduce the amount of time dedicated to formatting code, and also to help in applying the same coding standards in all notebooks. To apply automatic SQL formatting to a cell, you can select it from the cell context menu. This is only applicable to SQL code cells:

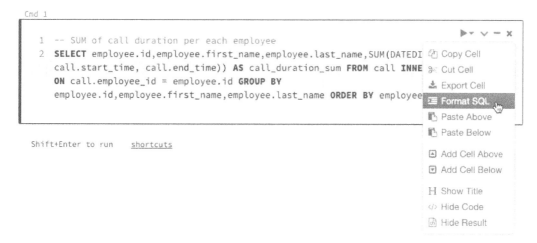

Figure 1.24 – Automatic formatting of SQL code

Applying the autoformatting of SQL code is a feature that can improve the readability of our code, and reduce possible mistakes due to bad formatting.

Exploring data management

In this section, we will dive into how to manage data in Azure Databricks in order to perform analytics, create ETL pipelines, train ML algorithms, and more. First, we will briefly describe types of data in Azure Databricks.

Databases and tables

In Azure Databricks, a database is composed of tables; table collections of structured data. Users can work with these tables, using all of the operations supported by Apache Spark DataFrames, and query tables using Spark API and Spark SQL.

These tables can be either global or local, accessible to all clusters. Global tables are stored in the Hive metastore, while local tables are not.

Tables can be populated using files in the DBFS or with data from all of the supported data sources.

Viewing databases and tables

Tables related to the cluster you are currently using can be viewed by clicking on the data icon button in the sidebar. The **Databases** folder will display the list of tables in each of the selected databases:

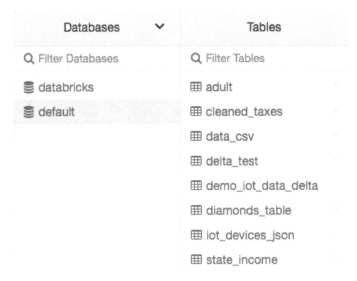

Figure 1.25 – Default tables

Users can select a different cluster by clicking on the drop-down icon at the top of the **Databases** folder and selecting the cluster:

Figure 1.26 – Selecting databases in a different cluster

We can have several queries on a cluster, each with its own filesystem. This is very important when we reference data in our notebooks.

Importing data

Local files can be uploaded to the Azure Databricks filesystem using the UI.

Data can be imported into Azure Databricks DBFS to be stored in the FileStore using the UI. To do this, you can either go to the **Upload Data** UI and select the files to be uploaded as well as the DBFS target directory:

Figure 1.27 – Uploading the data UI

Another option available to you for uploading data to a table is to use the **Create Table** UI, accessible in the **Import & Explore Data** box in the workspace:

Figure 1.28 – Creating a table UI in Import & Explore Data

For production environments, it is recommended to use the DBFS CLI, DBFS API, or the Databricks filesystem utilities (`dbutils.fs`).

Creating a table

Users can create tables either programmatically using SQL, or via the UI, which creates global tables. By clicking on the data icon button in the sidebar, you can select **Add Data** in the top-right corner of the **Databases** and **Tables** display:

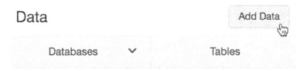

Figure 1.29 – Adding data to create a new table

After this, you will be prompted by a dialog box in which you can upload a file to create a new table, selecting the data source and cluster, the path to where it will be uploaded into the DBFS, and also be able to preview the table:

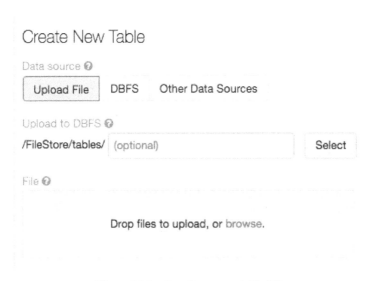

Figure 1.30 – Creating a new table UI

Creating tables through the UI or the **Add data** options are two of the many options that we have to ingest data into Azure Databricks.

Table details

Users can preview the contents of a table by clicking the name of the table in the **Tables** folder. This will show a view of the table where we can see the table schema and a sample of the data that is contained within:

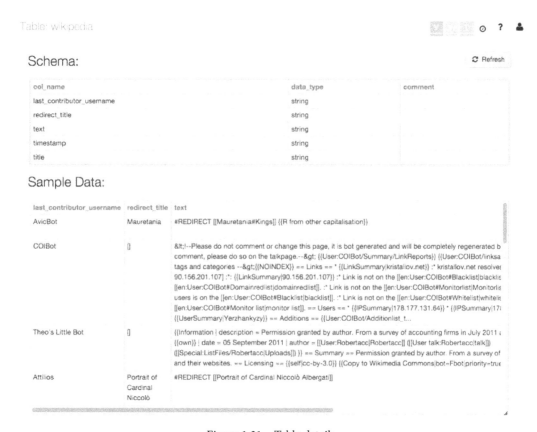

Figure 1.31 – Table details

These table details allow us to plan transformations in advance to fit data to our needs.

Exploring computation management

In this section, we will briefly describe how to manage Azure Databricks clusters, the computational backbone of all of our operations. We will describe how to display information on clusters, as well as how to edit, start, terminate, delete, and monitor logs.

Displaying clusters

To display the clusters in your workspace, click the clusters icon in the sidebar. You will see the **Cluster** page, which displays clusters in two tabs: **All-Purpose Clusters** and **Job Clusters**:

Figure 1.32 – Cluster details

On top of the common cluster information, **All-Purpose Clusters** displays information on the number of notebooks attached to them.

Actions such as terminate, restart, clone, permissions, and delete actions can be accessed at the far right of an **all-purpose** cluster:

Figure 1.33 – Actions on clusters

Cluster actions allow us to quickly operate in our clusters directly from our notebooks.

Starting a cluster

Apart from creating a new cluster, you can also start a previously terminated cluster. This lets you recreate a previously terminated cluster with its original configuration. Clusters can be started from the **Cluster** list, on the cluster detail page of the notebook in the cluster icon attached dropdown:

Quickstart Notebook (SQL)

▲ | ● 5.5 ML   ▾

Attached cluster:

● 5.5 ML
    DBR 5.5 ML | Spark 2.4.3 | Scala 2.11
    Detach Start Cluster Spark UI Driver Logs

Detach & Attach:

● 5.5
    DBR 5.5 | Spark 2.4.3 | Scala 2.11

Figure 1.34 – Starting a cluster from the notebook toolbar

You also have the option of using the API to programmatically start a cluster.

Each cluster is uniquely identified and when you start a terminated cluster, Azure Databricks automatically installs libraries and reattaches notebooks to it.

Terminating a cluster

To save resources, you can terminate a cluster. The configuration of a terminated cluster is stored so that it can be reused later on.

Clusters can be terminated manually or automatically following a specified period of inactivity:

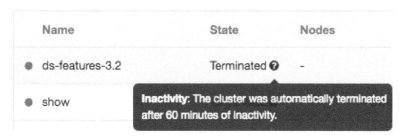

Figure 1.35 – A terminated cluster

It's good to bear in mind that inactive clusters will be terminated automatically.

Deleting a cluster

Deleting a cluster terminates the cluster and removes its configuration. Use this carefully because this action cannot be undone.

To delete a cluster, click the delete icon in the cluster actions on the **Job Clusters** or **All-Purpose Clusters** tab:

Figure 1.36 – Deleting a cluster from the Job Clusters tab

You can also invoke the permanent delete API endpoint to programmatically delete a cluster.

Cluster information

Detailed information on Spark jobs is displayed in the Spark UI, which can be accessed from the cluster list or the cluster details page. The Spark UI displays the cluster history for both active and terminated clusters:

Figure 1.37 – Cluster information

Cluster information allows us to have an insight into the progress of our process and identify any possible bottlenecks that could point us to possible optimization opportunities.

Cluster logs

Azure Databricks provides three kinds of logging of cluster-related activity:

- Cluster event logs for life cycle events, such as creation, termination, or configuration edits

- Apache Spark driver and worker logs, which are generally used for debugging

- Cluster init script logs, valuable for debugging init scripts

Azure Databricks provides cluster event logs with information on life cycle events that are manually or automatically triggered, such as creation and configuration edits. There are also logs for Apache Spark drivers and workers, as well cluster init script logs.

Events are stored for 60 days, which is comparable to other data retention times in Azure Databricks.

To view a cluster event log, click on the **Cluster** button at the sidebar, click on the cluster name, and then finally click on the **Event Log** tab:

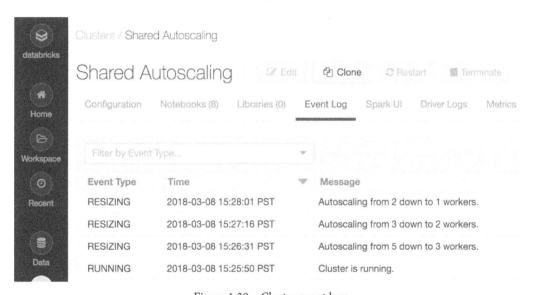

Figure 1.38 – Cluster event logs

Cluster events provide us with specific information on the actions that were taken on the cluster during the execution of our jobs.

Exploring authentication and authorization

Azure Databricks allows the user to perform access control to manage access to workspace objects, clusters, pools, and data tables. Admin users manage access control lists and also users with delegated permissions.

Clustering access control

By default, in Azure Databricks, all users can create or modify clusters. Before using cluster access control, an admin user must enable it. After this, there are two types of cluster permissions, which are as follows:

- The **Allow Cluster Creation** permission allows the creation of clusters.
- **Cluster-level permissions** allow you to manage clusters.

When cluster access control is enabled, only admins and users with **Can Manage** permissions can configure, create, terminate, or delete clusters.

Configuring cluster permissions

Cluster access control can be configured by clicking on the **cluster** button in the sidebar and, in the **Actions** options, selecting the **Permissions** button. This will prompt a permission dialog box where users can do the following:

- Apply granular access control to users and groups using the **Add Users** and **Groups** options.
- Manage granted access for users and groups.

These options are visible in *Figure 1.39*:

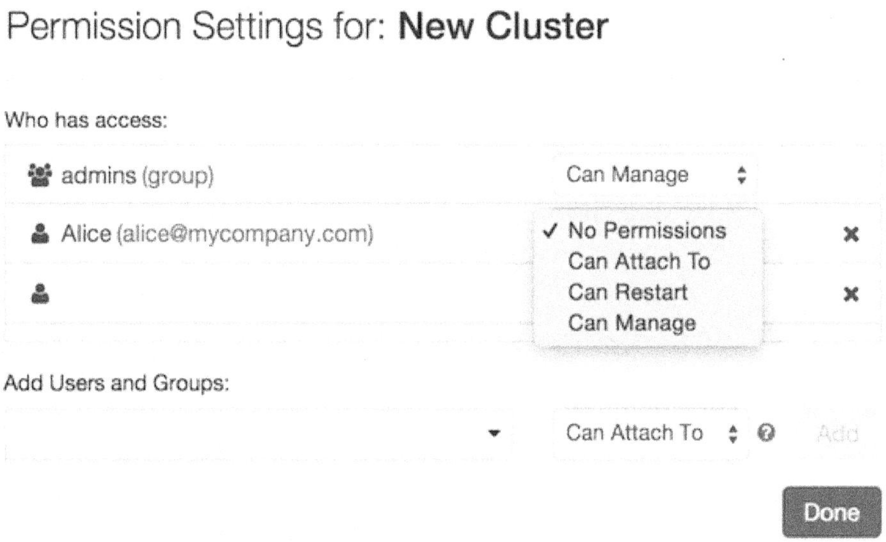

Figure 1.39 – Managing cluster permissions

Cluster permissions allow us to enforce fine-grained control over the computational resources used in our projects.

Folder permissions

Folders have five levels of permissions: **No Permissions**, **Read**, **Run**, **Edit**, and **Manage**. Any notebook or experiment will inherit the folder permissions that contain them.

Default folder permissions

Besides the current access control, these permissions are maintained:

- Objects in the **Shared** folder can be managed by anyone.
- Users can manage objects created by themselves.

When there is no workspace access control, users can only edit items in their **Workspace** folder.

With workspace access control enabled, the following permissions exist:

- Only admins can create items in the **Workspace** folder, but users can manage existing items.

- Permissions applied to a folder will be applied to the items it contains.

- Users keep having **Manage** permission to their home directories.

Understanding these permissions helps us to know in advance how possible changes in these policies could affect how users interact with the organization's data.

Notebook permissions

Notebooks have the same five permission levels as folders: **No Permissions**, **Read**, **Run**, **Edit**, and **Manage**.

Configuring notebook and folder permissions

Users can configure notebook permissions by clicking on the **Permissions** button in the notebook context bar. Select the folder and then click on **Permissions** from the drop-down menu:

Figure 1.40 – Notebook permissions

From there, you can grant permissions to users or groups as well as edit existing permissions:

Figure 1.41 – Access control on notebooks

Access control on notebooks can easily be applied in this way by selecting one of the options from the drop-down menu.

MLflow Model permissions

You can assign six permission levels to **MLflow Models** registered in the **MLflow Model Registry**: **No Permissions, Read, Edit, Manage Staging Versions, Manage Production Versions, and Manage**.

Default MLflow Model permissions

Besides the current workspace access control, these permissions are maintained:

- Models in the registry can be created by anyone.
- Administrators can manage any model in the registry.

When there is no workspace access control, users can manage any of the models in the registry.

With workspace access control enabled, the following permissions exist:

- Users can manage only the models they have created.
- Only administrators can manage models created by other users.

These options are applied to MLflow Models created in Azure Databricks.

Configuring MLflow Model permissions

One thing to keep in mind is that only administrators belong to the admins with the **Manage** permissions group, while the rest of the users belong to the **all users** group.

MLflow Model permissions can be modified by clicking on the model's icon in the sidebar, selecting the model name, clicking on the drop-down icon to the right of the model name, and finally selecting **Permissions**. This will show us a dialog box from which we can select specific users or groups and add specific permissions:

Figure 1.42 – MLflow permissions

You can update the permissions of a user or group by selecting the new permission from the **Permission** drop-down menu:

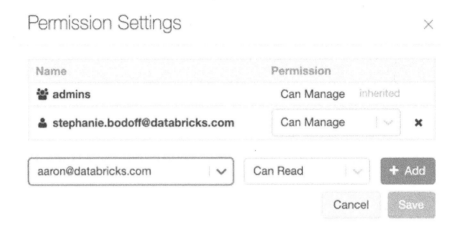

Figure 1.43 – MLflow access management

By selecting one of these options, we can control how MLflow experiments interact with our data and which users can create models that work with it.

Summary

In this chapter, we have tried to cover all the main aspects of how Azure Databricks works. Some of the things we have discovered include how notebooks can be created to execute code, how we can import data to use, how to create and manage clusters, and so on. This is important because when creating ETLs and ML experiments in Azure Databricks within an organization, aside from how to code the ETL in our notebooks, we will need to know how to manage the data and computational resources required, how to share assets, and how to manage the permissions of each one of them.

In the next chapter, we will apply this knowledge to explore in more detail how to create and manage the resources needed to work with data in Azure Databricks, and learn more about custom VNets and the different alternatives that we have in order to interact with them, either through the Azure Databricks UI or the CLI tool.

2
Creating an Azure Databricks Workspace

In this chapter, we will apply all the concepts we explored in *Chapter 1*, *Introduction to Azure Databricks*. We will create our first Azure Databricks workspace using the UI, and then explore the different possibilities of resource management through the Azure CLI, how to deploy these resources using the ARM template, and how we can integrate Azure Databricks within our virtual network using VNet injection.

In this chapter, we will discuss the following topics:

- Using the Azure portal UI
- Examining Azure Databricks authentication
- Working with VNets in Azure Databricks
- Azure Resource Manager templates
- Setting up the Azure Databricks CLI

We will first begin by creating our workspace from the Azure portal UI.

Technical requirements

The most important prerequisite for this chapter is to already have an Azure subscription with funds and permissions. Remember that this is a pay-as-you-go service, but nevertheless, you can create a free trial subscription; check out more information about this option in the Azure portal (`https://azure.microsoft.com/en-us/free/`).

Using the Azure portal UI

Let's start by setting up a new Databricks workspace through the Azure portal UI:

1. Log in to the Azure portal of your subscription and navigate to the Azure services ribbon.

2. Click on Azure Databricks:

Figure 2.1 – Creating an Azure Databricks service

3. This will lead you to the Azure Databricks default folder in which you will see all your resources listed. Click on Create new resource to create an Azure Databricks workspace environment:

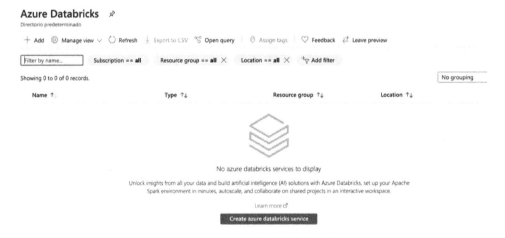

Figure 2.2 – Your Azure Databricks deployed resources

4. Once you click on Create azure databricks service, you will have to fill in a couple of details regarding the workspace you are creating. The settings will look something like this:

- Workspace name
- Subscription
- Resource group
- Location
- Pricing Tier
- Deploy Azure Databricks workspace in your Virtual Network (Preview)

The name of the workspace can be whatever you like, but it is always good to pick a name that is simple and references the use that it will have. For example, these environments can be acceptance or production, and each one would be its own type of subscription. The resource group can be created at that moment and must logically cover all the resources that have the same purpose.

The location to deploy your resources is also important because it must be close to the point of service or to the other deployed resources with which it must work together.

> **Note**
>
> Regarding the pricing tier: there are three options, which are Standard, Premium, and Trial. One of the biggest differences is that Premium allows you to apply access control. This makes it a common choice within corporate environments.

One of the options presented is that you can choose to deploy in your own dedicated virtual network. In this step, we will select No, but we will talk later in the chapter about VNet and how it can be a good security practice to have all your Azure resources under the same network security policy.

5. Finally, you can create the resource, and the workspace will be automatically configured and deployed:

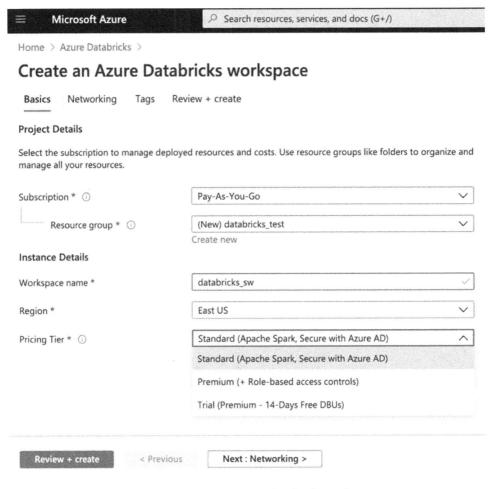

Figure 2.3 – Creating an Azure Databricks workspace

Now that we have successfully created our workspace, we can access it through the Workspace UI of the Azure CLI.

Accessing the Workspace UI

Once your Azure Databricks resource has been deployed successfully, you can access the workspace. You can directly access it by clicking on the Launch Workspace button at the center of your resource in the Azure portal, or you can access it by using the URL that is shown in the center-left corner of your resource. This URL will take you to the workspace that you will access using Azure AD sign-in:

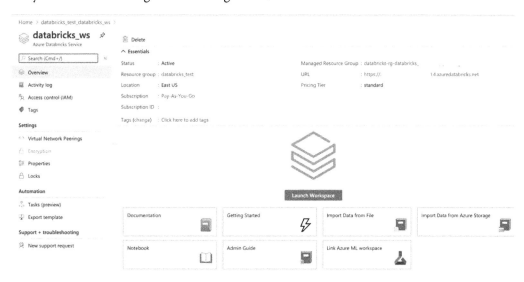

Figure 2.4 – The Azure Databricks service main panel

Once you can access the workspace, you will see different options to access your assets. You can either use the left-hand ribbon of options, or the center options to open notebooks, create new clusters, new jobs, and so on. The entire list of options accessible in the main left-hand ribbon is as follows:

- Azure Databricks
- Home
- Workspace
- Recent assets
- Data
- Clusters

- Jobs
- Search:

We can see the UI here:

Figure 2.5 – The Azure Databricks workspace

Once we have created our workspace environment, before we can focus on creating our first notebook, first you need to create a cluster. The cluster is necessary because it's there that the code in our notebooks will run. After creating a cluster, the next step will be to create a new notebook, which is the collaborative script on which Azure Databricks is based.

Configuring an Azure Databricks cluster

As we have mentioned before, a cluster is where the code in our notebooks is run. It can be thought of as a set of VMs running in the background and balancing the load from our computations. This resource is billable when it's on, so remember to terminate it when not in use.

Let's click on the Clusters icon on the left ribbon. This will open a window with the options to create a new cluster. The only necessary field is the name of the cluster, the rest of the fields have default values that we could leave as they are, but it's good that we look at them more closely. The options that we have are as follows:

- Cluster Mode: We can choose from a Standard to a High Concurrency cluster.

- Pool: This is the number of standby compute machines ready and waiting to be used if the computing needs increase during the process. It reduces spin-up time for the cluster.

- Databricks Runtime Version: The version of Scala or Spark running in the cluster. Most available options now just allow Python 3.

- Autopilot Options: Enables autoscaling of the resource up and down and terminates following a predefined number of minutes of inaction.

The final configuration is the type of driver and worker we will use. The driver is responsible for coordinating and distributing the job to the workers. More workers, or better workers or drivers, will increase the computing performance of the cluster, but will impact the cost of these resources, so be sure to check the pricing first:

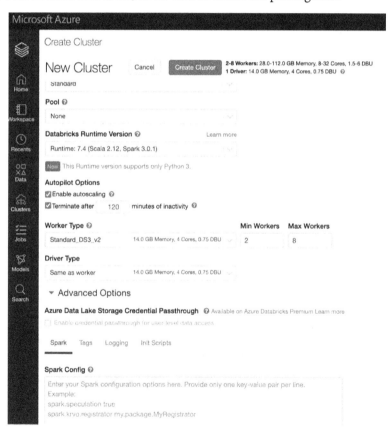

Figure 2.6 – Creating a cluster

After the cluster has been created successfully, we can create a new notebook and attach it to the new cluster and start running code on it.

Creating a new notebook

Let's go back to the workspace and from there, we have two options to create a new notebook:

1. The first option is to create it directly from the home workspace by clicking on the New Notebook link under Common Tasks. This will open a dialog box from where we can select the default language of the notebook and the cluster to which it will be attached:

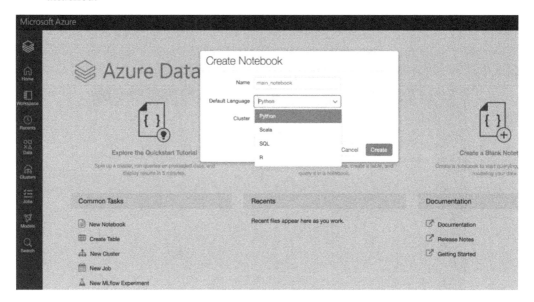

Figure 2.7 – Creating a notebook

The other option that we have is to create the notebook directly in a specific directory:

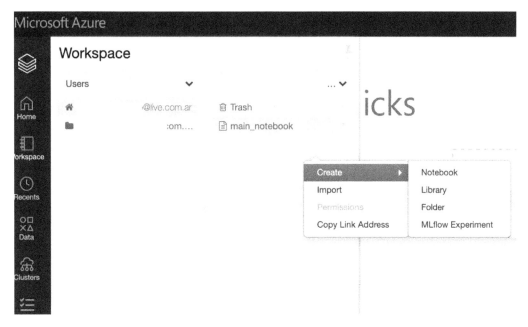

Figure 2.8 – Creating a notebook in a specific location

2. Once the notebook has been created, we attach it to a cluster, and we can start creating cells and running code in them:

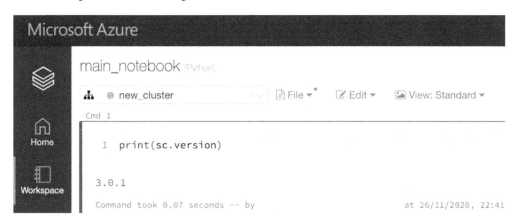

Figure 2.9 – Notebook attached to a cluster

You can use Python because it's the default language of the notebook we just created, but if you want to use a different language, for example, Scala, remember you can use the magic %scala command. To execute the code in a cell, you can use the shortcuts *Ctrl + Enter* or *Shift + Enter*. If the cluster that the notebook is attached to is not running, you will be prompted to confirm that you want it running again.

We are now almost ready to handle data, but to do this, we have a couple of things that we have to bear in mind.

There are three different ways in which we handle data from Azure Storage from PySpark:

- Using a WASB file path, which is a way to specify the path to a file in the Windows Azure storage blob.

- Using a mount point on worker nodes with the FS protocol. We can use the magic %fs command to specify commands such as ls, and we can use the files stored there, using file paths such as dbfs:/mnt//<containername>/<partialPath>.

- Using a mount point and requesting files using a regular file path: /dbfs/ mnt/<containername>/<partialPath>.

We can also upload data directly by creating a table with the Create Table UI:

Figure 2.10 – Creating a new Table UI

Now, we also have the option to upload files into the notebooks directly:

Figure 2.11 – Uploading data to the notebook

3. The other nice option we have is that Databricks already has some datasets to start with! We can use fs utils to explore this; by way of an example, we can use one of the coronavirus datasets:

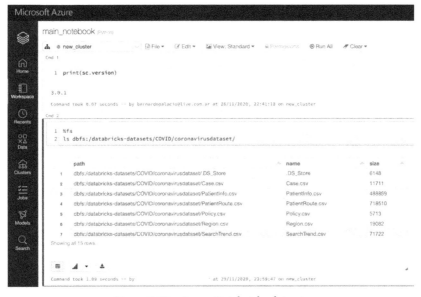

Figure 2.12 – Azure Databricks datasets

4. After that, we are ready to use the data. We will load it as a Spark DataFrame. The code to do this is as follows:

```
file_path = "dbfs:/databricks-datasets/COVID/
coronavirusdataset/PatientInfo.csv"

df = spark.read.format("csv").load(file_path,header =
"true",inferSchema = "true")

display(df)
```

The result is visible in the following screenshot:

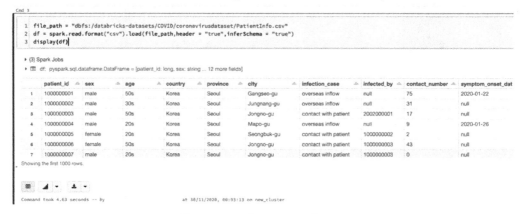

Figure 2.13 – Coronavirus dataset

5. After that we can run some statistics on the data frame:

```
df.printSchema()
df.describe().show()
df.head(5)
```

You can see the result in the following screenshot:

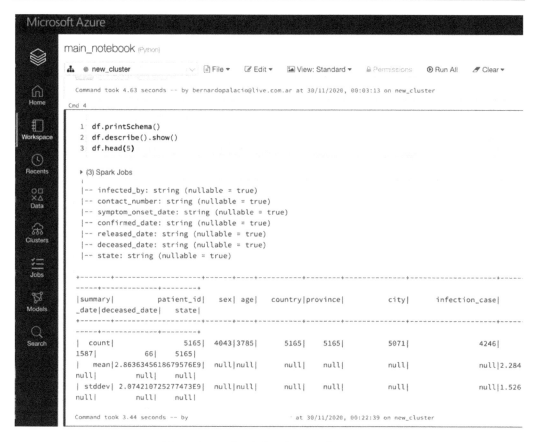

Figure 2.14 – Spark data frame information

So far, we have seen how to create an Azure workspace and create a notebook in which we can run code and work with data. In the following parts of the chapter, we will dive deeper into how we can control these resources through the Azure CLI, deploying with ARM templates, and VNet injection.

Now we will see how we can share these notebooks through Azure Databricks authentication.

Examining Azure Databricks authentication

In Azure Databricks, authentication is effected through our Azure AD account, which, in some cases, can be linked to our Microsoft account. Subscriptions such as Premium allow us to manage access to our assets through access control in a more detailed manner.

Access control

If we share the URL of our workspace with a user to collaborate, first we must grant access to that user. To do this, we could either grant them Owner or Contributor roles of the asset we want to share or do this in Admin Console:

1. You can access Admin Console by clicking on the resource name icon in the top-right corner:

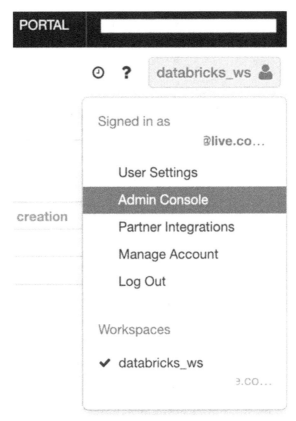

Figure 2.15 – Admin Console

2. After that, users can be added by clicking on the Add User button and then selecting the role you would like that user to have. You can create internal groups and apply a more detailed control over folder and workspace assets:

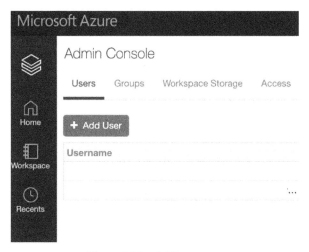

Figure 2.16 – Adding a user

Users can be added to our workspace to work collaboratively with our notebooks and data.

Working with VNets in Azure Databricks

Azure Databricks can be deployed within a custom virtual network. This is called VNet injection and is very important from a security perspective. When we deploy with default settings, inbound traffic is closed, but outbound traffic is open without restrictions. When we use VNet injection and we deploy directly to a custom virtual network, we can apply the same security policies around all our Azure Services, to meet compliance and security requirements.

In case you are working in data science or exploratory environments, it's good to leave the outbound traffic open to be able to download packages and libraries for Python, R, and Maven, and Ubuntu packages also.

As we have mentioned before, Azure Databricks works on two planes of service. The first is the control page, which we use through the Databricks API to work with workspace assets. The second is the data plane where the clusters are deployed. It is this second plane that is deployed to your customized virtual network. The benefits of this are as follows:

- A secure connection to other Azure services, such as Azure Storage, through service endpoints

- A connection to on-premises data sources to be used in Azure Databricks

- Connecting Azure Databricks to a network virtual appliance to monitor and control outbound traffic

- Configuring Azure Databricks to use custom DNS
- Configuring network security group (NSG) rules
- Deploying Azure Databricks clusters in your existing virtual network

This is possible because of Azure Databricks delegation, which allows for joint control of a subnet. The owner of the network allows resources to be deployed in a specific subnet and exerts control by specifying conditions for this delegation to happen and add or modify security policies.

Virtual network requirements

The VNet where you intend to deploy your Azure Databricks workspace should meet the following criteria in order to work:

- It should have the same location as the Azure Databricks resource we are creating.
- The VNet must belong to the same subscription as the Azure Databricks resource.
- The VNet should have private and public dedicated subnets for Azure Databricks, to allow communication with the control plane.
- Address space: A CIDR block between /16 – /24 for the virtual network, and a CIDR block up to /26 for the private and public subnets.

Virtual networks allow us to enforce the same policies in all our Azure resources.

Deploying to your own VNet

We will show the basics of how to deploy a cluster to a specific virtual network. Azure Databricks will create two subnets within this network a network security groups using the CIDR ranges provided by you, whitelist inbound and outbound traffic, and deploy once the virtual network has been updated.

Deployment requirements

To deploy Azure Databricks, you will first have to configure your virtual networks in the following way:

- The virtual network can be an existing one or a newly created one but, as mentioned before, the location and subscription must be the same as the Azure Databricks workspaces that you plan to create.
- A CIDR range between /16 – /24 is required for the virtual network.

Configuring the virtual network

While we were *creating an Azure Databricks resource*, we were prompted with an options tab called Networking. In this tab, we will see the options to deploy to our own virtual network if we select to deploy to our own virtual network:

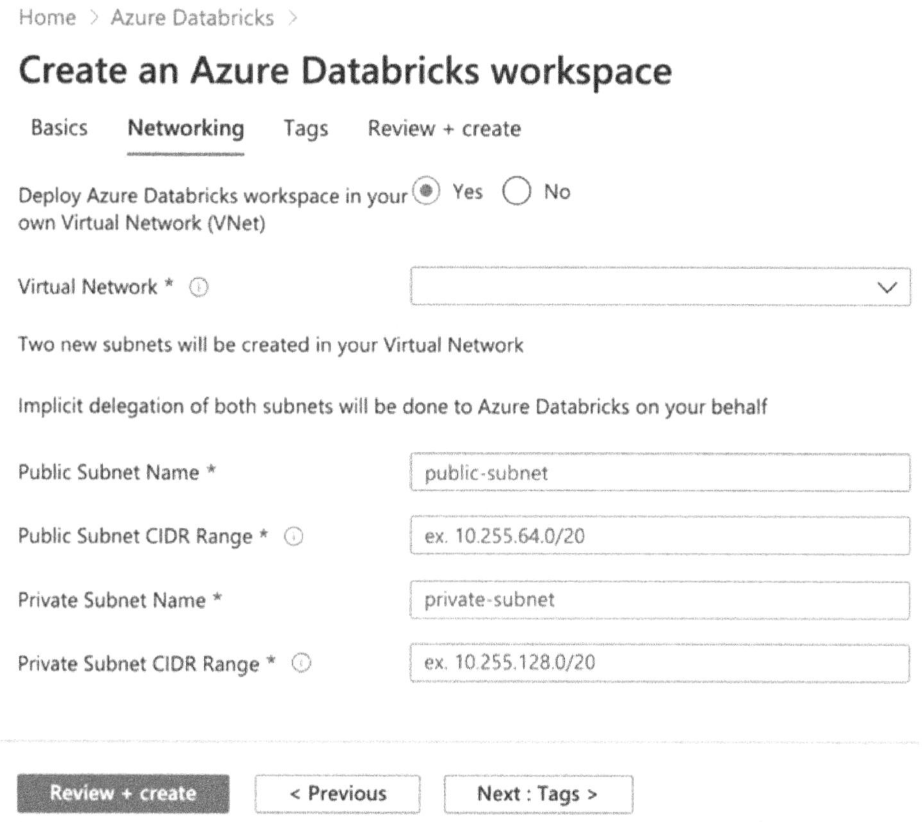

Figure 2.17 – Azure Databricks in a custom virtual network

From here, we can select the network we want to use, the subnets public to the internet, the private subnets, and the CIDR range. Once this is set, you can click on Review + create and wait for your Azure Databricks to be deployed within your network.

We can specify the configuration of our virtual networks in much more detail using Azure Resource Manager (ARM). These are templates that we can use to deploy services to Azure. With these templates, we can specify whether we want to use existing subnets, use an existing security group, and more. In the next section, we will take a deeper look into ARM templates for Azure Databricks.

Azure Resource Manager templates

ARM templates are infrastructure as code and allow us to deploy resources automatically in an agile manner. These templates are JSON files that define infrastructure and configuration in a declarative way, specifying resources and properties. We can deploy several resources as a single resource, and modify existing configurations. Just like code, it can be stored in a repository and versioned, and anyone can run the code to deploy similar environments.

ARM templates are then passed to the ARM API, which deploys the specified resources. These can include virtual networks, VMs, or an Azure Databricks workspace.

These templates have two modes of operation, which are Complete or Incremental mode. When we deploy in Complete mode, this deletes any objects that are not specified in the template and the resource group that is being deployed to. Incremental deployment adds additional resources to the existing ones.

The limitation of these templates is that they do not deploy code along with the resources.

It is also important to mention that if we are deploying resources using an ARM template, it is good to first test it in another test subscription in Azure. That can help us validate whether the template is deployable through the Azure CLI and avoid, for example, deploying in Complete mode and then failing to create new resources because of a limitation on your Azure subscription quota. We will see how to do that in the next section.

Creating an Azure Databricks workspace with an ARM template

We will see how we can create an Azure Databricks workspace using an ARM template and then validate that it was correctly deployed. To do this, we will use as an example the ARM template from Azure Quickstart Templates.

As usual, to create an Azure Databricks workspace, we require an Azure subscription:

1. Go to the Azure portal and, in the search tab, look for Deployment of Custom Template. From here, we can select to build our own template or use one from the GitHub Quickstart Template. We will select this option and look for 101-databricks-workspace, but there are options for using VNet injection and managed keys:

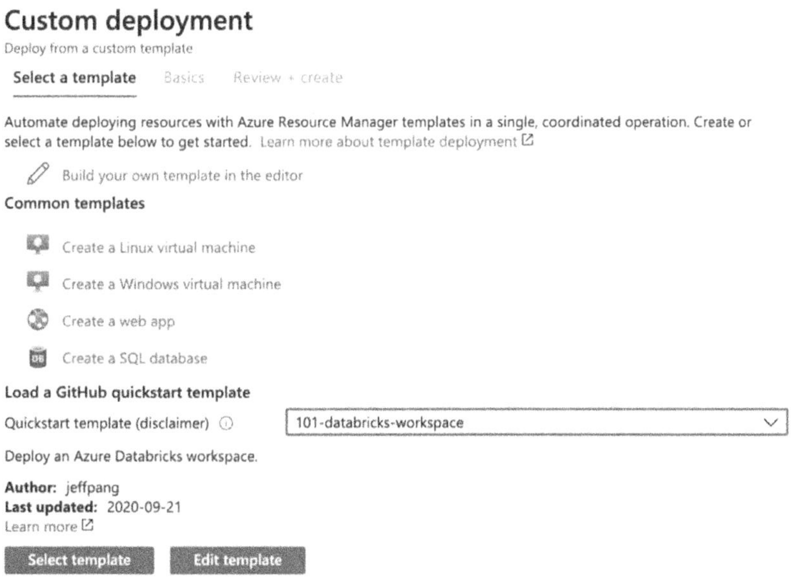

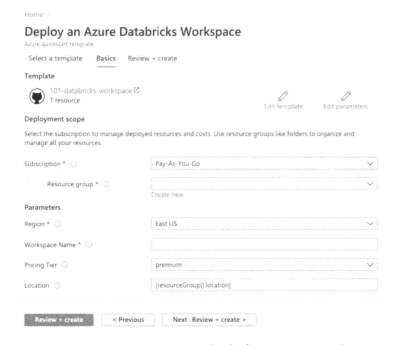

Figure 2.18 – Custom deployment

2. After selecting the template, you need to provide the required values and then select Review + create, and finally Create:

Figure 2.19 – Creating Azure Databricks from an ARM template

As we have probably experienced before, the creation of the workspace takes some time. If the deployment fails, the workspace will be in a failed state and we will need to delete it and create a new one. When this workspace is deleted, the managed resource group and any other deployed resource will be deleted too.

Reviewing deployed resources

You can check that the Azure Databricks workspace has been successfully deployed in the Azure portal, or use the following Azure CLI script to list the resource:

```
echo "Enter your Azure Databricks workspace name:" &&
read databricksWorkspaceName &&
echo "Enter the resource group where the Azure Databricks
workspace exists:" &&
read resourcegroupName &&
az databricks workspace show -g $resourcegroupName -n
$databricksWorkspaceName
```

Cleaning up resources

If you are following these as a tutorial and you are not planning to use the resources created previously, you can delete them using the next Azure CLI script:

```
echo "Enter the Resource Group name:" &&
read resourceGroupName &&
az group delete --name $resourceGroupName &&
echo "Press [ENTER] to continue ..."
```

Let's talk a little bit more about the Azure CLI. It's a tool that we can use to interact with our resources in Azure Databricks. In the next section, we will describe in more detail how we can use it to manage our workspace assets.

Setting up the Azure Databricks CLI

Azure Databricks comes with a CLI tool that allows us to manage our resources. It's built on top of the Azure Databricks API and allows you to access the workspace, jobs, clusters, libraries, and more. This is an open source project hosted on GitHub.

The Azure Databricks CLI is based on Python 3 and is installed through the following pip command:

```
pip3 install databricks-cli
```

You can confirm that the installation was successful by checking the version. If the installation was successful, you will see as a result the current version of the Azure Databricks CLI:

```
databricks --version
```

It's good to bear in mind that using the Databricks CLI with firewall-enabled storage containers is not possible and, in that case, it is recommended to use Databricks Connect or AZ storage.

To be able to install the Azure CLI, you will need to have Python 3 already installed and added to the path of the environment you will be working on.

Authentication through an access token

Let's set up the CLI:

1. To be able to use CLI commands, we will have to authenticate ourselves by setting an Azure Databricks personal access token. This token is generated in your Azure Databricks profile, in the access token section:

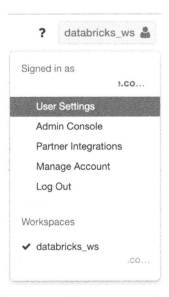

Figure 2.20 – User account settings

2. After this, we click on the Generate New Token button in the top-left corner. This
    will require us to complete a comment on what this access token will be used for
    and establish a lifetime period for our token:

Figure 2.21 – Generate New Token

3. Next, you will see the generated token. Copy and store it in a safe place as this will
    be the only time that the token will be shown to you. Once created, the token will be
    listed in the Access Token list in your User Settings tab in your account.

4. Now that we have the token, we can use it to authenticate ourselves in the Azure
    CLI. To do this, in our terminal, we can write the following command:

```
databricks configure --token
```

We will need to provide the host and the token. The host is the URL of the Azure
Databricks resource we are trying to access.

Authentication using an Azure AD token

The other option that we have for configuring the CLI is to use an Azure AD token:

1. You can generate this access token in the Azure portal and once this is done, we can
    store it as an environment variable called DATABRICKS_AAD_TOKEN:

```
export DATABRICKS_AAD_TOKEN=<azure-ad-token>
```

2. After we have set the environment variable, we can run the next command to
    configure the Databricks CLI using the AD token. This command will require you
    to provide the URL of the Azure Databricks resource we are trying to access, which
    will have the format adb-<workspace-id>.<random-number>.azuredatabricks.net:

```
databricks configure --aad-token.
```

3. After completing all the required fields prompted to you, your access credentials will be stored in a file, ~/.databrickscfg. This file should contain entries such as the following:

```
host = https://<databricks-instance>
token = <azure-ad-token>
```

In versions of CLI from 0.8.1 and above, the path of this file can be changed by setting the environment variable, DATABRICKS_CONFIG_FILE, to the path we want to.

Validating the installation

Once we have provided these parameters, we will be able to use the Azure Databricks CLI to interact with our resources:

1. We can test this by running the next command, which will show us the assets in our user folder in the workspace:

```
databricks workspace ls /Users/<your_username>
```

As a result, you will get a list of all the assets in your user folder.

2. We can always get the help documentation related to the available commands by running the following command:

```
databricks <resource> -h
```

For example, this resource could be the DBFS, for which we will need to fill the resource in brackets with fs, leaving the command as follows:

```
databricks fs -h
```

Workspace CLI

As we have seen before, the workspace is the root folder for our Azure Databricks environment and is where the notebooks and folders exist. We can manage these assets through the Databricks CLI. For example, we could use the CLI tool to copy a Python script to a specific folder in our workspace:

1. The command to do this would be as follows:

```
databricks workspace import <source_path>/script.py
<target_path> -l PYTHON
```

Here, the -l option is used to indicate the language of the file we have uploaded.

2. After this, we can validate that the file has been correctly imported by running the following command:

```
databricks workspace list /<target_path>
```

As we have seen before, this will show us all the files and folders in our target path.

In addition to this, we have commands that allow us to delete, export, and create folders, and more. Some of them are mentioned next, but you can use the -h commands to see all of them in the CLI tool.

Using the CLI to explore the workplace

Here are some more ways in which you can get information about your workspace using the Azure CLI:

- *List workspace files*:

```
databricks workspace ls /Users/example@databricks.com
```

- *Import a local directory of notebooks*: The databricks workspace import_dir command allows us to import all the files and directories from a local filesystem to the workspace. It's worth mentioning that only directories and files with the extensions of .scala, .py, .sql, .r, and. R are imported and when this is done, these extensions are stripped from the notebook name.

  We can choose to delete all existing notebooks at the target path by adding the flag -o:

```
databricks workspace import_dir . /Users/example@
databricks.com/example
```

- *Export a workspace folder to the local filesystem*: You can export a folder of notebooks from the workspace to a local directory. The command to do this is as follows:

```
databricks workspace export_dir /Users/example@
databricks.com/example .
```

Clusters CLI

Clusters can be listed and created, and information about them can be retrieved from the Clusters CLI. If you want to create a cluster, you will have to supply a JSON configuration file for that cluster. You can read more about this file in the Azure Databricks documentation:

- To list all the commands available to work with clusters, run the following command in your terminal:

```
databricks clusters -h
```

- We can also list all the runtime versions of our clusters:

```
databricks clusters spark-versions
```

- And we can also list all the node types that exist in our resource:

```
databricks clusters list-node-types
```

Jobs CLI

The Jobs CLI allows for creating, editing, and running jobs in Azure Databricks. Jobs are more complicated than other APIs available for Databricks because many of the commands require passing JSON configuration files, as we have seen with clusters creation. As we are used to, the full list of commands can be accessed using the -h option in the Jobs CLI:

```
databricks jobs -h
```

Listing and finding jobs

The databricks jobs list command has two output formats, JSON and TABLE. The TABLE format is outputted by default and returns a two-column table (job ID, job name). The Jobs CLI allows us to use regular expressions to find one specific job.

To find a job by name, run the following command:

```
databricks jobs list | grep "JOB_NAME"
```

Groups API

The Groups API allows for the management of groups of users, whose members can be added and removed from it. Groups can be created, deleted, and listed. The following command will list all members in the admins group:

```
databricks groups list-members --group-name admins
```

There are five permission levels for Databricks, which are, ordered hierarchically: No Permissions, Read, Run, Edit, and Manage. Read allows cells to be viewed and comments to be made on notebooks. Run adds to this the possibility of attaching and detaching notebooks from clusters and running notebooks. Edit adds to the previous permissions the possibility to edit cells. Manage can perform all actions and change the permissions of others.

The Databricks CLI from Azure Cloud Shell

One of the benefits that we have in Azure is that we can use the Databricks CLI directly from the Azure portal:

1. The button next to the search bar is the Cloud Shell icon:

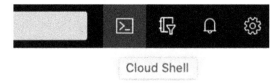

Figure 2.22 – Azure Cloud Shell

2. If we click on it, we will be prompted with the option of Bash or PowerShell. To use the CLI, we must select Bash:

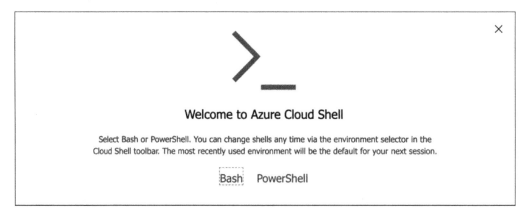

Figure 2.23 – Azure Cloud Shell Bash

3. If you have no storage created, you will be prompted with a message that says that we require storage to persist files and we can create it there directly.

4. Once the storage has been created, we have access to the Azure Cloud Shell in which we can create a virtual environment and install the Databricks CLI in the same way that we would do in a local terminal.

Once you complete these steps, you can start using the Databricks CLI from Azure Cloud Shell.

Summary

In this chapter, we explored the possibilities of creating an Azure Databricks service either through the UI or the ARM templates, and explored the options in terms of enforcing access control on resources. We also reviewed the different authentication methods, the use of VNets to have a consistent approach when dealing with access policies throughout Azure resources, and how we use the Databricks CLI to create and manage clusters, jobs, and other assets. This knowledge will allow us to efficiently deploy the required resources to work with data in Azure Databricks while maintaining control over how these assets access and transform data.

In the next chapter, we will apply this and the previous concepts to run more advanced notebooks, create ETLs, data science experiments, and more. We'll start with ETL pipelines in *Chapter 3, Creating ETL Operations with Azure Databricks*.

Section 2: Data Pipelines with Databricks

This section will help you work with data and learn how to manage different sources of data, transform it, and use it in the interactions of the different options that Azure has in order to create entire ETL pipelines.

This section contains the following chapters:

3
Creating ETL Operations with Azure Databricks

In this chapter, we will learn how to set up different connections to use external sources of data such as **Simple Storage Service** (**S3**), set up our Azure Storage account, and use Azure Databricks notebooks to create **extract, transform, and load** (**ETL**) operations that clean and transform data. We will leverage **Azure Data Factory** (**ADF**), and finally, we will look at an example of designing an ETL operation that is event-driven. By exploring the sections in this chapter, you will be able to have a high-level understanding of how data can be loaded from external sources and then transformed into data pipelines, constructed and orchestrated using Azure Databricks. Let's start with a brief overview of **Azure Data Lake Storage Gen2** (**ADLS Gen2**) and how to use it in Azure Databricks.

In this chapter, we will look into the following topics:

- Using ADLS Gen2
- Using S3 with Azure Databricks
- Using Azure Blob storage with Azure Databricks
- Transforming and cleaning data

- Orchestrating jobs with Azure Databricks
- Scheduling jobs with Azure Databricks

Technical requirements

To follow the examples given in this chapter, you need to have the following:

- An Azure subscription
- An Azure Databricks notebook attached to a running container
- An **Amazon Web Services** (**AWS**) S3 bucket with the files that we want to access, along with access credentials with the proper permissions to read files in the bucket

Using ADLS Gen2

To persist data in Azure Databricks, we need a data lake. We will use ADLS Gen2, so our first step is to set up an account. This will allow us to store permanent data and use it to run ETL pipelines, get analytics, or use it to build **machine learning** (**ML**) models.

Setting up a basic ADLS Gen2 data lake

To set up an ADLS Gen2 subscription, we first need to create a new resource in our Azure portal. To do this, follow these next steps:

1. Search for **Storage accounts** and select **Create a new Storage account**.
2. Attach it to a resource group, set up a name, set the **Account kind** field to **StorageV2 (general-purpose v2)** and, finally, set **Replication** to **Locally-redundant storage (LRS)**, as illustrated in the following screenshot:

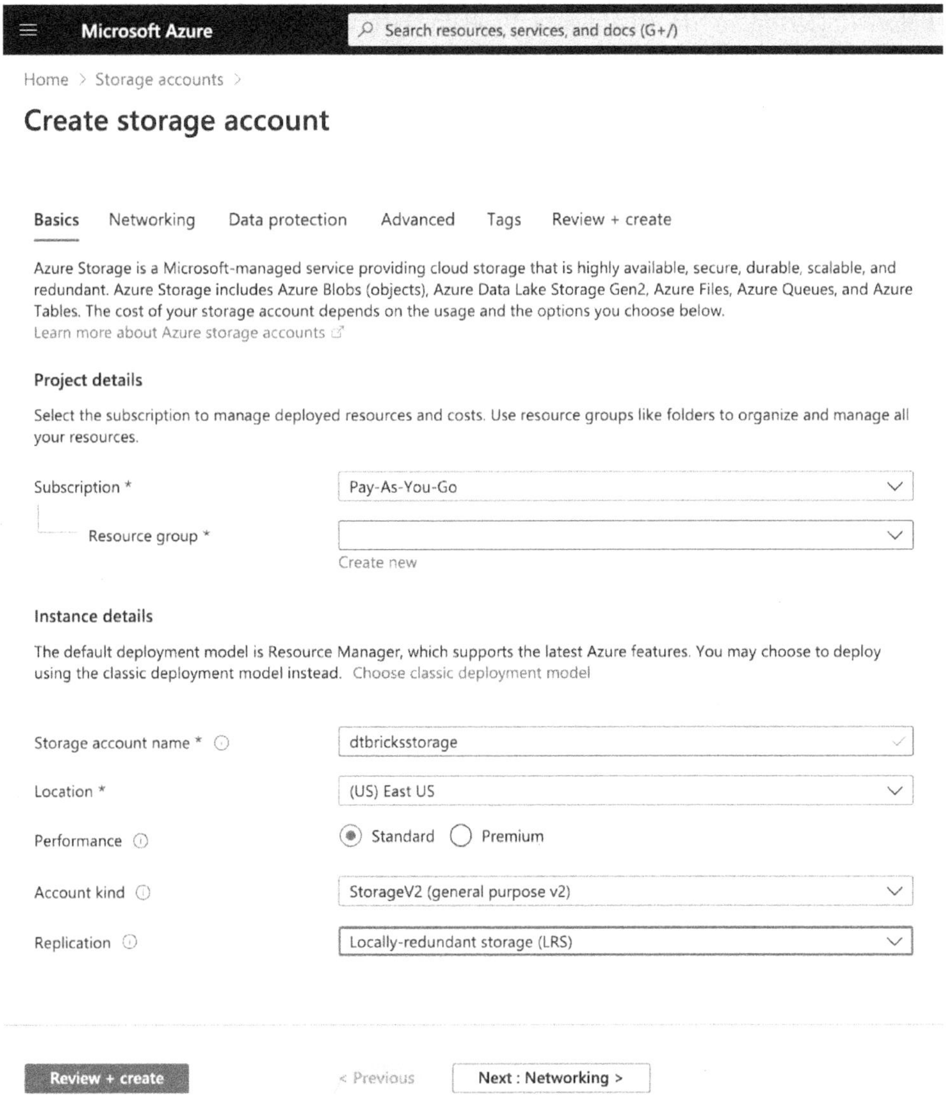

Figure 3.1 – Creating an ADLS Gen2 subscription

3. Before finalizing, in the **Advanced** tab, set the **Hierarchical namespace** option to **Enabled** so that we can use ADLS Gen2 from our notebooks. The following screenshot illustrates this:

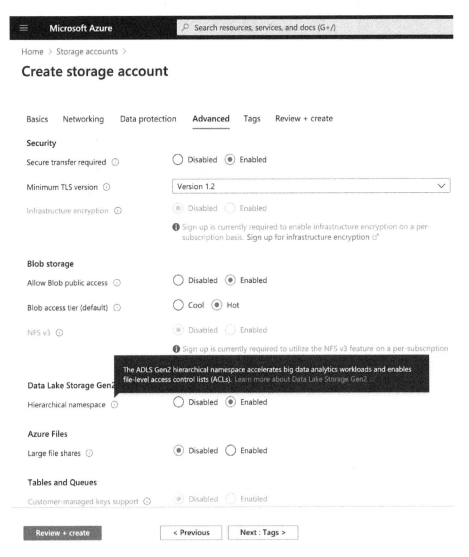

Figure 3.2 – Enabling the Hierarchical namespace option

4. After this, you can click on the **Review + create** button. Your resource will be validated and you can then start the deployment, which usually takes a couple of minutes. Finally, we will have our ADLS Gen2 storage account set up. Now, we can upload files to it, for which we will use **Storage Explorer**.

Next, we'll add in the data.

Uploading data to ADLS Gen2

To upload the file that we will use in our demo, search for **Storage Explorer** in the Azure portal, and then follow these steps:

1. Select a subscription from the left tab—this will show all the containers related to that subscription. Right-click on the container in the ADLS Gen2 storage account and create a new filesystem that will contain our data. The process is illustrated in the following screenshot:

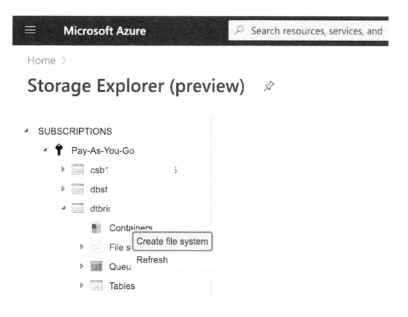

Figure 3.3 – Creating a filesystem

2. After the filesystem has been created, select the **Upload** button from the options bar at the top of the main panel, as illustrated in the following screenshot. This will prompt us with a message to download **Storage Explorer**, which is a program that allows us to upload files stored locally and manage our storage account:

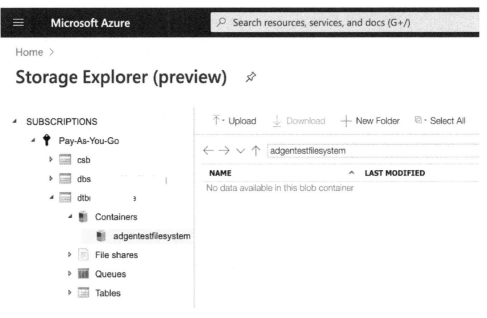

Figure 3.4 – Selecting the Upload button to download Storage Explorer

3. Follow the steps to install **Storage Explorer** on your local machine.

4. Once it is installed, log in to your Azure account and navigate to the left tab until you find the subscription in which we have created our ADLS Gen2 account. There, we will see the filesystem that we have created, and from there, we can upload local files using the **Upload** button in the top-left bar, as illustrated in the following screenshot:

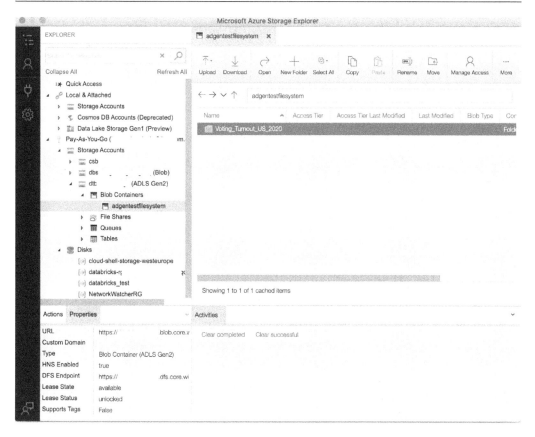

Figure 3.5 – Azure Storage Explorer

The data that we will be using is the `2020 US General Election Turnout rates` data, which comprises the turnout rates for the voting-eligible population in the 2020 **United States** (**US**) election between Donald Trump and Joe Biden. The voter turnout is the percentage of eligible voters who voted in an election, and the US government has been publishing this data since the beginning of elections in the early days of US independence.

The data is from the Census Bureau's **Current Population Survey** (**CPS**); the *November Voting and Registration Supplement* can be downloaded from the following link: `https://data.world/government/vep-turnout`.

Now that we have set up our storage account and uploaded the files, we need to access them from our Azure Databricks notebook. Let's see how we can do this.

Accessing ADLS Gen2 from Azure Databricks

According to the documentation, we have three ways to access ADLS Gen2 from Azure Databricks, outlined as follows:

- Using ADLS Gen2 access keys on the notebook
- Mounting the ADLS Gen2 filesystem into the Databricks file system at a cluster level
- Using a service principal to access data without mounting

We will use the first option, which is the most direct, therefore we need the access key from the storage account that we want to access.

It is worth mentioning that in production environments, it is best practice to save these keys in Azure Key Vault and then use Azure Databricks to link them, and use these keys as environment variables in our notebooks. The steps are as follows:

1. To get the keys to access our ADLS Gen2, select the **Storage account** option from the resource panel in the Azure portal and then go to **Access keys** in the left tab. Copy **key1**, which is the key that we will be using to access the data from the notebook. The process is illustrated in the following screenshot:

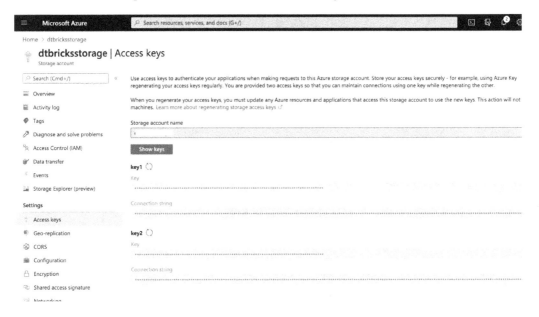

Figure 3.6 – ADLS Gen2 access keys

2. In our notebook, we can set the connection to our ADLS Gen2 account. Replace your storage account name in the connection string and paste **key1** from the account into the following command to set up the connection:

```
spark.conf.set(
   "fs.azure.account.key.<storage-account-name>.dfs.core.
windows.net",
   <key-1>
)
```

In this manner, the credentials will be stored in the session configuration and we won't need to reenter them unless the clusters get restarted.

3. After this has been run successfully, we can use dbutils to list the directory where our data is stored. Replace the filesystem name and storage account name in the connection string of the following command to list all the files in the directory where the data is stored:

```
dbutils.fs.ls("abfss://<file-system-name>@<storage-
account-name>.dfs.core.windows.net/<directory-name>")
```

We will see, as a result, a list of files in that directory. We can save it, allowing us to load it into our notebook.

Loading data from ADLS Gen2

Now that we have stored the keys in Azure Key Vault, we can load that data into the **ADFS** from ADLS Gen2, as follows:

1. After you have listed all the files in your directory, copy and save the path of the file that we uploaded to a variable. We will name this variable file_location, as can be seen in the following code snippet:

```
file_location = "<your-file-path>"
```

2. After that, we can read it and load it into a Spark dataframe and display it by running the following code:

```
df = spark.read.format("csv").option("inferSchema",
"true").option("header", "true").option("delimiter",",").
load(file_location)
display(df)
```

3. With the following command, we can see the schema of data types that were inferred, and if they're not correct, we can make the necessary corrections:

```
df.printSchema()
```

Now, we can use data stored in ADLS Gen2 directly from Azure Databricks notebooks, which will allow us to use it as a data sink or data lake while building data pipelines.

Using S3 with Azure Databricks

To work with S3, we will suppose that you already have a bucket in AWS S3 that contains the objects we want to access and that you have already set up an access key ID and an access secret key. We will store those access keys in the notebook as variables and use them to access our files in S3 directly, using Spark dataframes.

Connecting to S3

To make a connection to S3, store your AWS credentials in two variables named aws_ access_key_id and aws_secret_access_key into the Hadoop configuration environment. Use the following commands, which assume that you have already saved your credentials as variables:

```
sc._jsc.hadoopConfiguration().set("fs.s3n.awsAccessKeyId", aws_
access_key_id)
sc._jsc.hadoopConfiguration().set("fs.s3n.awsSecretAccessKey",
aws_secret_access_key)
```

After this is set, we can access our bucket directly by reading the location of our file.

Loading data into a Spark DataFrame

After our credentials have been saved in the Hadoop environment, we can use a **Spark dataframe** to directly extract data from S3 and start performing transformation and visualizations.

In the following lines of code, we will read the file stored in the S3 bucket and load it into a Spark dataframe to finally display it. PySpark will use the credentials that we have stored in the Hadoop configuration previously:

```
my_bucket = "<your-bucket-name>"
my_file = "<your-file>"
df = spark.read.csv(f"s3://{my_bucket}/{my_file}",
```

```
                header=True, inferSchema=True)
display(df)
```

This produces the following output:

Figure 3.7 – Data fetched from S3

We have formatted the string connection to the file that we want to read in our bucket and successfully loaded it into a Spark dataframe. We will see now that the way to connect to Azure Blob storage is very similar.

Using Azure Blob storage with Azure Databricks

In the same way that we can access objects stored in AWS S3, we can access objects in **Azure Blob storage**. Both options allow us to have a redundant data storage that can be accessed from anywhere. Their differences lie in the tools that they will be used with and certain characteristics that may make them more suitable to use in a certain project. Azure Blob storage is more cost-efficient and has high redundancy, while S3 is extensively used by several organizations and has a small learning curve. We will see how to set up an Azure Blob storage account, upload the file that we were using, and read it from our notebook.

Setting up Azure Blob storage

The first step is to create an Azure Blob storage account, as follows:

1. Search for **Storage account** in the Azure portal and select **Create a new storage account**.

2. In the **Create storage account** options, fill out the details and set the **Replication** option to **Locally-redundant storage (LRS),** and for the **Account kind** option select **Blob storage**, as illustrated in the following screenshot:

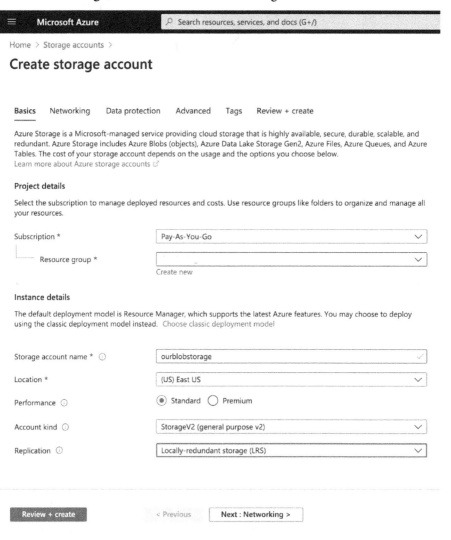

Figure 3.8 – Creating an Azure Blob storage account

3. After this, select the **Review + create** button, wait for the validation to be successful, and deploy the storage account.

Uploading files and access keys

Next, we'll upload our files and access keys. To do this, follow these steps:

1. After our storage account has been deployed, navigate to the resource and in the left tab, select **Containers** and create a new one, as illustrated in the following screenshot:

Figure 3.9 – Creating a container in Azure Blob storage

2. After we have selected a name, we can create a new container.

3. Once it's created, we can upload files by clicking the **Upload** button from the top bar, which will deploy a tab at the right where we can select a file to upload, as illustrated in the following screenshot:

Figure 3.10 – Uploading a file to Azure Blob storage

4. Once our file has been uploaded, go to **Access keys** in the left panel and copy **key1**, which we will use to read the files from our Azure Databricks notebook. The process is illustrated in the following screenshot:

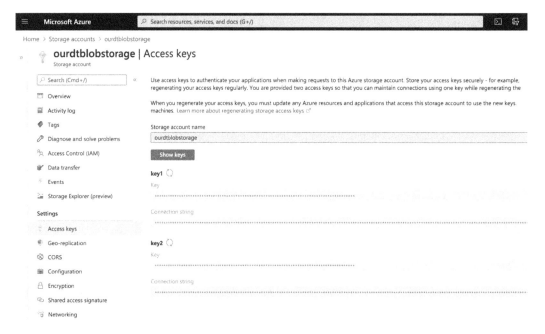

Figure 3.11 – Azure Blob storage access keys

Once we have set up the Azure Blob storage account, uploaded the file to it, and copied the access key, we can set up our connection from the Azure Databricks notebook.

Setting up the connection to Azure Blob storage

After our files have been uploaded to the Blob storage and we have our access keys, we can connect our Azure Databricks notebooks. To do this, we will use the following command:

```
spark.conf.set(
  "fs.azure.account.key.<your-storage-account-name>.blob.core.
windows.net",
  "<your-storage-account-access-key>")
```

Thus, following the previous example, the configuration command would look like this:

```
spark.conf.set("fs.azure.account.keyourdtblobstorage.blob.core.
windows.net","<key-1>")
```

This command sets the connection to our connection to Azure Blob storage. Remember to change the connection string with the name of your storage account and with the storage access key (remember that this is **key1** from the resource in the Azure portal).

After this is set, we can use the `dbutils` tool to list our container in the Blob storage with the following command:

```
dbutils.fs.ls("wasbs://<your-container-name>@<your-storage-
account-name>.blob.core.windows.net/<your-directory-name>")
```

In the next section, we'll look at how to transform and clean the data we want to use.

Transforming and cleaning data

After our data has been loaded into a Spark dataframe, we can manipulate it in different ways. We can directly manipulate our Spark dataframe or save the data to a table, and use **Structured Query Language (SQL)** statements to perform queries, **data definition language (DDL)**, **data manipulation language (DML)**, and more.

You will need to have the `Voting_Turnout_US_2020` dataset loaded into a Spark dataframe.

Spark data frames

A Spark data frame is a tabular collection of data organized in rows with named columns, which in turn have their own data types. All this information is stored as metadata that we can access using `displaySchema` in order to display the data types of each column or display the actual data, or `describe` in order to view the statistical summary of the data. One of its characteristics is that it is able to handle big amounts of data thanks to its distributed nature.

We can perform transformations such as selecting rows and columns, accessing values stored in cells by name or by number, filtering, and more thanks to the PySpark **application programming interface** (**API**).

We will use these transformations in combination with SQL statements to transform and persist the data in our file.

Querying using SQL

We will use the voting turnout election dataset that we have used before. We will create a view of the data and use SQL to query it.

Creating views

We can create views using the data in the Spark Dataframe to be able to query it using SQL, as follows:

1. To create a view that we will call voter_turnout, we can use the following command:

    ```
    df.createOrReplaceTempView("voter_turnout")
    ```

2. After this, we can use a SELECT command to get every record in the view that we have just created, as follows:

Figure 3.12 – Creating a temporary view from a Spark dataframe

Now that we have created a view of the data, we can query it using SQL commands.

3. We will select everything, and then we will filter using a WHERE clause to get all the results from the state of Arizona, as illustrated in the following screenshot:

Figure 3.13 – Using SQL statements to query data

Creating a temporary view of the data is helpful when we want to run experiments or tests, but after the cluster has been restarted, we will lose the view. To persist this data, we can create a permanent table.

Creating tables in a data lake

Permanent tables allow us to have persistent data that can be queried by several users and help us when the data needs to be frequently accessed. When we create a table, what we are doing is declaring metadata in the Hive metastore, which is the place where all our data is stored. The voting turnout file that we will use to create the table is stored in ADLS Gen2, which is a kind of blob storage, but it might as well be in the Azure DBFS, which is the storage that is created when we set up our workspace. The steps are as follows:

1. Before creating our table, we will create a database that will contain it. We will call the voting_data database, and it can be created using the following command:

   ```
   %sql
   CREATE DATABASE voting_data
   ```

   If everything turns out well, we should see OK as the output of the cell in which we run the command.

2. After the database has been created, we can define a table named voting_turnout_2020, which will be constructed using a **comma-separated values (CSV)** file that we have uploaded to ADLS Gen2. Remember that the file path we will use is the one we obtained when running the list directory command using dbutils previously. That said, we can create a permanent table using the following command:

   ```
   %sql
   CREATE TABLE IF NOT EXISTS voting_data.voting_
   ```

```
turnout_2020
USING CSV
LOCATION 'abfss://<your-filesystem>@<you-storage-
account>.dfs.core.windows.net/Voting_Turnout_US_2020/2020
November General Election - Turnout Rates.csv'
```

This gives us the following output:

Figure 3.14 – Creating a database and a table from a CSV file

3. After this is done, we can double-check that everything went well by clicking on
    the **Data** button in the left ribbon of our Azure Databricks workspace. We will see
    the recently created database and the table it contains, as illustrated in the following
    screenshot:

Figure 3.15 – The created table in Azure DBFS

This would allow us to, again, run SQL queries on our table, but with the difference that the table is now permanent. The table can be seen in the following screenshot:

Figure 3.16 – Querying data from a permanent table

4. One of the issues that we can see here is that the creation of the table from the CSV file failed to recognize the first rows as headers. The solution for this is simple, because we have an option in SQL in which we can state that our data has headers. To solve this, we will first need to drop the table using the following command:

```
%sql
DROP TABLE voting_data.voting_turnout_2020
```

5. And then, we need to rerun our CREATE TABLE command, but with an added option at the end to clarify that our file has headers, as illustrated in the following code snippet:

```
%sql
CREATE TABLE IF NOT EXISTS voting_data.voting_
turnout_2020
USING CSV
LOCATION 'abfss://adgentestfilesystem@dtbricksstorage.
dfs.core.windows.net/Voting_Turnout_US_2020/2020 November
General Election - Turnout Rates.csv'
OPTIONS (header "true", inferSchema "true")
```

6. After this, selecting all the records from our table will show us that we were successful, and our data now has headers, as can be seen in the following screenshot:

Figure 3.17 – Fixing headers in a table

7. If we want to check whether Spark was able to correctly infer the information schema of our file, we can run a DESCRIBE command on our table. This, in our case, would be the following command:

```
%sql
DESCRIBE voting_data.voting_turnout_2020
```

The output of this would be the data types on each one of the columns of our table, as illustrated in the following screenshot:

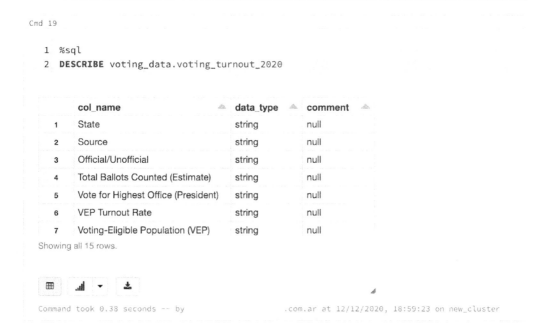

Figure 3.18 – Describing the schema of a table

That's covered the SQL queries, but we're not finished with the transforming and cleaning section yet.

Writing back table data to Azure Data Lake

After we have performed the necessary transformations to our data, we might want to write it back to ADLS Gen2. To do this, we can take the following steps:

1. Let's suppose we have filtered the results to have only entries for which the result is still Unofficial. We will execute the following command, which will copy the table into a Spark dataframe:

```
from pyspark.sql.functions import col
df_filtered = spark.table('voting_data.voting_
turnout_2020')
```

2. After this, we can apply the filtering, which can be done using the PySpark col function, and finally display the result, as follows:

```
# Filter all records that are not "Unofficial"
df_filtered = df_filtered.filter(col("Official/
Unofficial") == "Unofficial")
display(df_filtered)
```

This gives us the following output:

Figure 3.19 – Filtering rows in a Spark dataframe

3. One thing to bear in mind is that if we try to directly export this dataframe to ADLS Gen2, we will encounter an error because the column names have invalid characters on them. We will quickly rename the columns with names that have these characters replaced, using regular expressions. The code to do this is shown in the following snippet:

```
import re
oldColumns = df_filtered.schema.names
newColumns = [re.sub(r'\W', '', i) for i in oldColumns]
df_filtered = df_filtered.toDF(*newColumns)
display(df_filtered)
```

In the following screenshot, we can see that the column names have changed, and they now only have letters on them:

Figure 3.20 – Renaming Spark dataframe columns

4. After this, and to improve the efficiency of the storage, we can convert the data frame to Parquet format and store it in a location that we specify in ADLS Gen2. This can be done in the following way:

```
#declare data lake path where we want to write the data
target_folder_path = 'abfss://adgentestfilesystem@
dtbricksstorage.dfs.core.windows.net/Voting_Turnout_
Filtered/'
#write as parquet data
df_filtered.write.format("parquet").save(target_folder_
path)
```

Once we have executed this on a cell, we can confirm in Azure **Storage Explorer** that a new Parquet file has been created in the new folder that we have specified, as illustrated in the following screenshot:

Figure 3.21 – Created data shown in Storage Explorer

Now that our data has been uploaded, we can use it in jobs that can be scheduled and orchestrated with ADF.

Orchestrating jobs with Azure Databricks

Until now, we have been able to use data stored in either an S3 bucket or Azure Blob storage, transform it using PySpark or SQL, and then persist the transformed data into a table. Now, the question is: *Which methods do we have to integrate this into a complete ETL?* One of the options that we have is to use ADF to integrate our Azure Databricks notebook as one of the steps in our data architecture.

In the next example, we will use ADF in order to trigger our notebook by directly passing the name of the file that contains the data we want to process and use this to update our voting turnout table. For this, you will require the following:

- An Azure subscription
- An Azure Databricks notebook attached to a running container
- The `Voting_Turnout_US_2020` dataset loaded into a Spark dataframe

ADF

ADF is the Azure cloud platform for the integration of serverless data transformation and aggregation processes. It can integrate many services and provides a codeless **user interface** (**UI**) in order to create our ETLs.

Creating an ADF resource

The first step is to create an ADF resource. To do this, follow these steps:

1. In the Azure portal, search for **Azure Data Factory** and select **Create Resource**.

2. Select a name for our ADF resource and attach it to the resource group in which you're using Azure Databricks, as illustrated in the following screenshot:

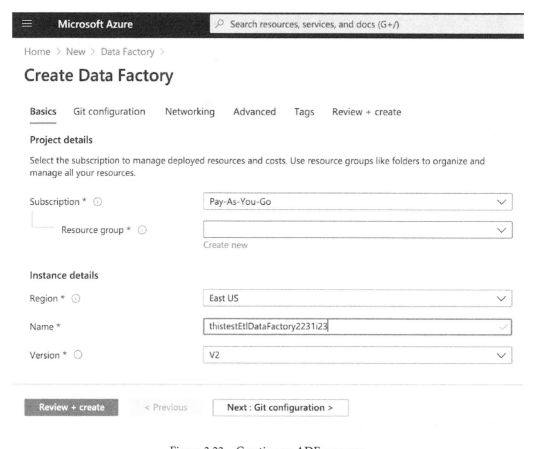

Figure 3.22 – Creating an ADF resource

3. Before creating the resource, go to the **Git configuration** tab and select **Configure Git later**, as illustrated in the following screenshot:

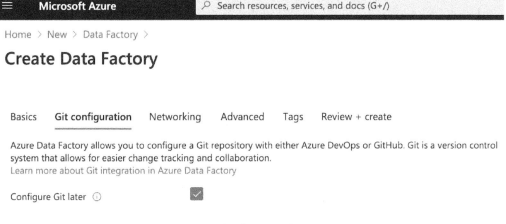

Figure 3.23 – Configure Git later option

4. After this is set up, we can click on the **Review + create** button, wait for it to be validated, and deploy it.

Creating an ETL in ADF

The steps are as follows:

1. Once it's deployed, go to the ADF resource and click on **Author & Monitor** in the main panel, as illustrated in the following screenshot:

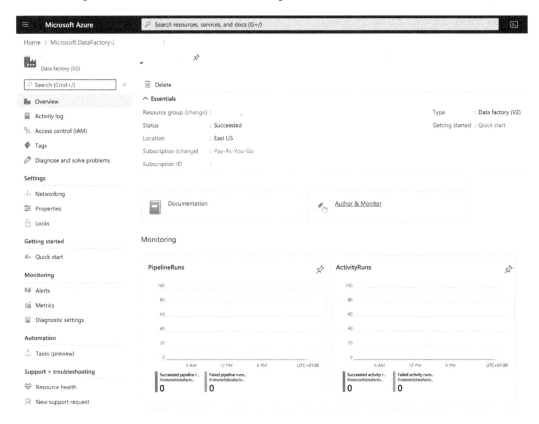

Figure 3.24 – Going to Author & Monitor in ADF

2. This will take us to the **Author & Monitor** panel, from which we can create and monitor ETL pipelines. On the left tab, select the **Author** button to create our ETL pipeline, as illustrated in the following screenshot:

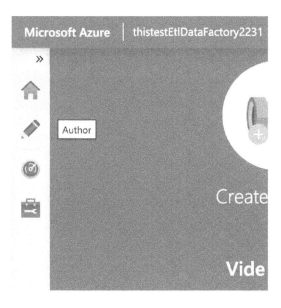

Figure 3.25 – Going to the Author button in the left tab

3. This will take us to the canvas UI, in which we can create our pipelines. Select
    Pipelines from the panel at the left of the window and set up a name and a
    description.

4. Then, click on **Search** for the Databricks panel at the left of the canvas, which will
    show our notebook in the canvas, as illustrated in the following screenshot:

Figure 3.26 – Creating a pipeline and adding a Databricks service

5. Select the Azure Databricks notebook in the canvas, and from the options at the bottom of the canvas, click on the **+ New** button on **Databricks linked service** to link our Azure Databricks workspace to ADF. In order to do this, we will need an authentication token, as illustrated in the following screenshot:

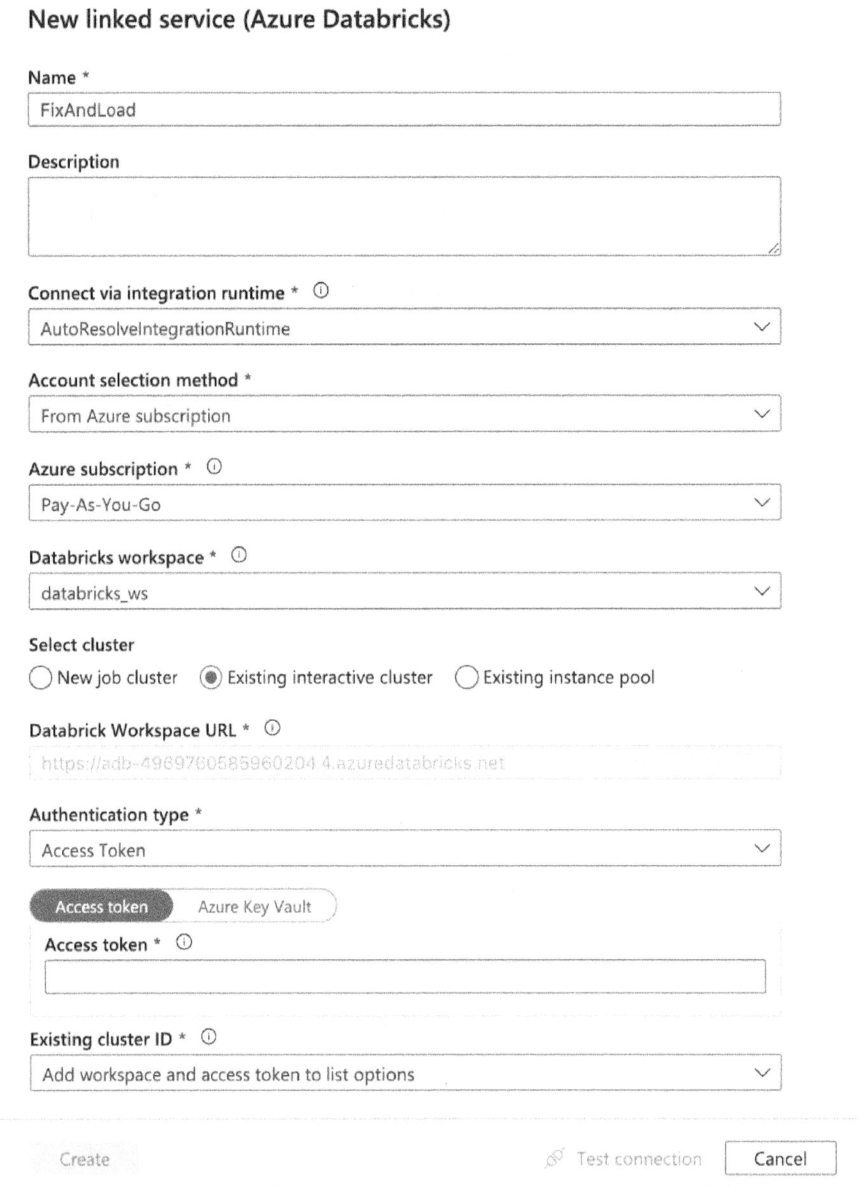

New linked service (Azure Databricks)

Name *

> FixAndLoad

Description

Connect via integration runtime * ⓘ

> AutoResolveIntegrationRuntime

Account selection method *

> From Azure subscription

Azure subscription * ⓘ

> Pay-As-You-Go

Databricks workspace * ⓘ

> databricks_ws

Select cluster

◯ New job cluster   ⦿ Existing interactive cluster   ◯ Existing instance pool

Databrick Workspace URL * ⓘ

> https://adb-4969760585960204.4.azuredatabricks.net

Authentication type *

> Access Token

⬤ Access token   Azure Key Vault

Access token * ⓘ

Existing cluster ID * ⓘ

> Add workspace and access token to list options

Create   ⌀ Test connection   Cancel

Figure 3.27 – Linking the Azure Databricks service with an access token

6. To generate the token, go to the Azure Databricks workspace and click on the **User Settings** option in your profile to take you to the screen shown in the following screenshot. Once there, generate a new token and paste it into the ADF **Link new service** dialog box, and finally, click on the **Create** button:

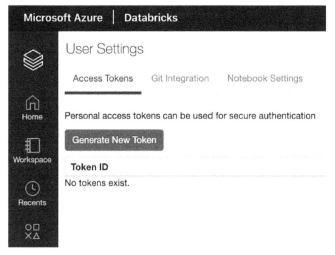

Figure 3.28 – Creating an Azure Databricks access token

7. Now, we can select our Azure Databricks workspace from the **Linked services** list and test the connection. If everything went well, the connection will be successful, as illustrated in the following screenshot:

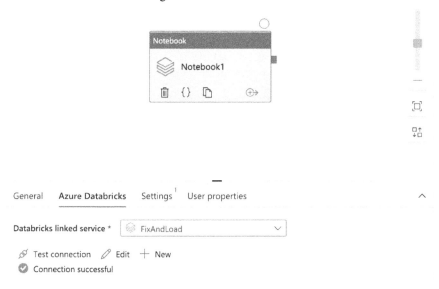

Figure 3.29 – Testing the Azure Databricks service connection

8. Once our Azure Databricks notebook has been correctly set up in our pipeline, we must establish a connection between it and ADF. We will use a variable that we will use when the ETL is triggered, and this will be the file that we want to process and persist on a table.

    If we click in the blank space of the canvas, at the bottom we will see a **Parameters** option. We will create a new parameter, which will be a string named input_ file, and have as the default value the location of the file that we want to process, as illustrated in the following screenshot:

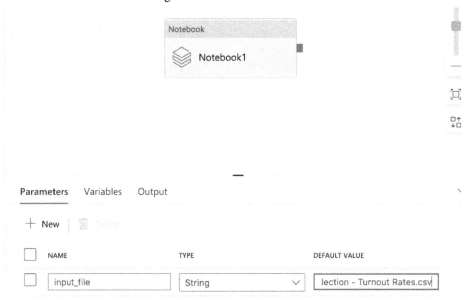

Figure 3.30 – Creating a parameter for the ETL pipeline

9. After that, click on the Azure Databricks notebook on the canvas and go to the **Settings** tab. There, we will browse and set up the notebook that we will use and add a base parameter that will point to the variable we have set up, as illustrated in the following screenshot:

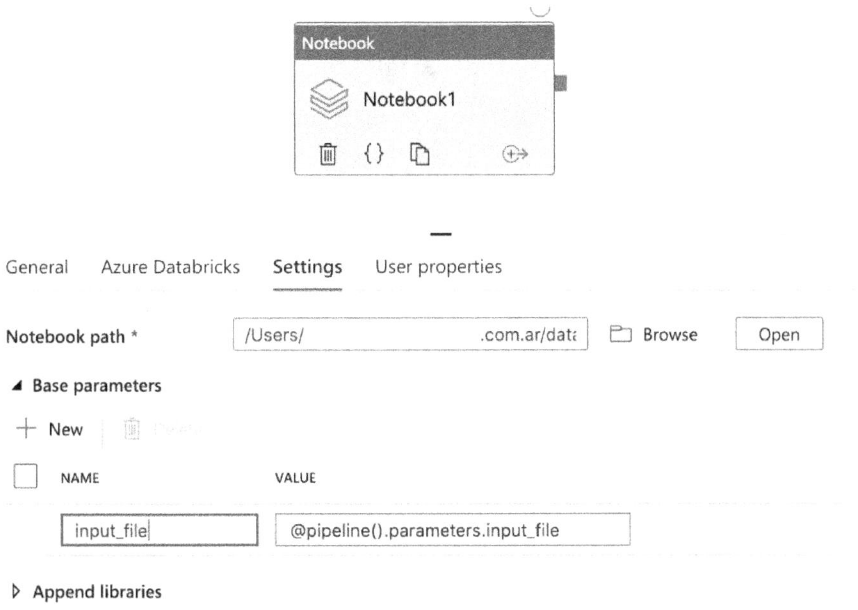

Figure 3.31 – Using the parameter as a variable in the Azure Databricks job

10. Until this point, we haven't discussed how this variable will be consumed in the notebook. To do this, we will add a piece of code into our notebook. We will specify that this variable is the name of the file that we need to process and will use it to format the full path of the file to be read. The code that we will use is shown in the following snippet:

```
# Creating widgets for leveraging parameters
dbutils.widgets.text("input", "","")
input_file = dbutils.widgets.get("input_file")
print ("Param -\'input':")
print (input_file)
```

This way, we are using dbutils to create a widget that will look for a variable named input_file, which is the one we have specified in ADF.

11. We will use this variable to format the path of the file that we will read and process in our notebook. The way in which we do this in our notebook is shown in the following code snippet:

```
#set the data lake file location:
file_location = f'abfss://<file-system-name>@<storage-
account-name>.dfs.core.windows.net/{input_file}'

#read in the data to dataframe df
df = spark.read.format("csv").option("inferSchema",
"true").option("header", "true").option("delimiter",",").
load(file_location)
```

12. After this step, we will apply the transformations that we consider necessary, and finally drop the existing table and save our transformed dataframe as a permanent table. The drop of the table can be done in SQL, using the following code:

```
%sql
DROP TABLE voting_data.voting_turnout_2020
And then we can save our spark dataframe as a table.
df.write.format("parquet").saveAsTable("voting_data.
voting_turnout_2020")
```

13. After our notebook is ready, as well as our ADF ETL, we can trigger this by clicking on the **Add Trigger** button, and then **Trigger Now**. ADF will prompt a dialog box in which we can specify the file and directory as the variable that we have established as a parameter previously. After this is set, we can click on **Run** and our pipeline will be triggered, as illustrated in the following screenshot:

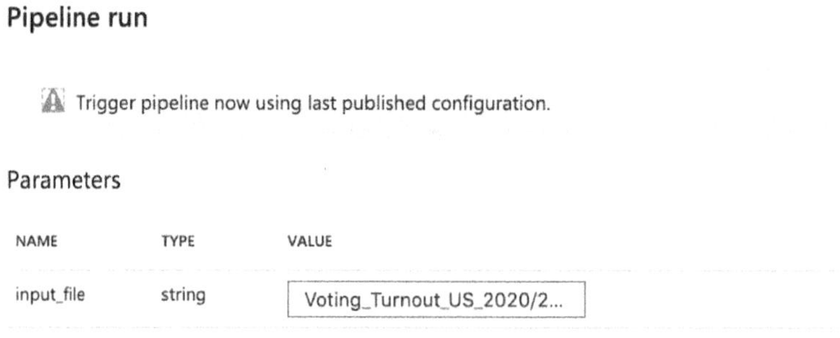

Pipeline run

⚠ Trigger pipeline now using last published configuration.

Parameters

NAME	TYPE	VALUE
input_file	string	Voting_Turnout_US_2020/2...

Figure 3.32 – Running the pipeline

14. We can check the status of our pipeline by clicking on the **Monitor** tab in ADF, which will show us the status of our pipeline, as illustrated in the following screenshot:

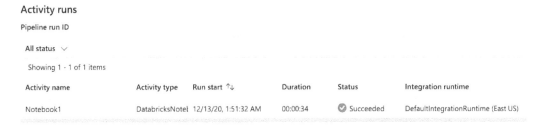

Figure 3.33 – Seeing the pipeline logs in the Monitor tab

15. As we can see, our pipeline was successfully executed, which means the file that we have specified was read, processed, and used to update our table.

Next, we'll see how to use Databricks to schedule jobs.

Scheduling jobs with Azure Databricks

If we already know that the file we want to process will be delivered to the blob storage, we can directly schedule the notebook to run periodically. To do this, we can use Azure Databricks jobs, which is an easy way to schedule the runs of our notebooks. We will suppose now that the file path of the file we will consume is fixed.

Scheduling a notebook as a job

The steps are as follows:

1. To schedule a new job, click on the **Jobs** tab in the left ribbon of our workspace and then click on **Create Job**, as illustrated in the following screenshot:

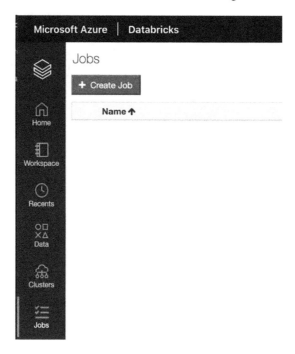

Figure 3.34 – Creating an Azure Databricks job

2. After this, the rest is quite straightforward. We will be required to specify which notebook we will use, set up an execution schedule, and specify the computational resources we will use to execute the job. In this case, we have chosen to run the job in an existing cluster, but we can create a dedicated cluster for specific executions. We select the notebook that will execute the job, as illustrated in the following screenshot:

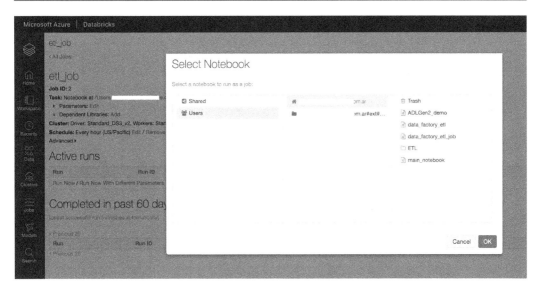

Figure 3.35 – Selecting a notebook

3. Next, we schedule the job to run at a certain time, with a specific starting date and frequency, as illustrated in the following screenshot:

Figure 3.36 – Scheduling the job

Once everything is set, we can run our job and wait for the execution to be completed to check on the status.

Job logs

If everything went well, we will see the status marked as **Succeeded**, as illustrated in the following screenshot. Otherwise, it will be marked as **Failed**, and we will need to check the logs of that specific run to figure out what went wrong:

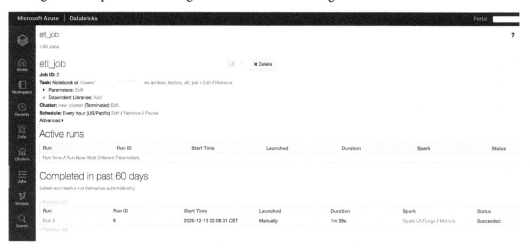

Figure 3.37 – Ongoing and completed runs for the job

Luckily, our job was successful, which means our file will be processed and our table will be updated every hour.

Summary

In this chapter, we have applied the concepts learned previously to create and connect to Azure Databricks resources to store our data, such as ADLS Gen2, AWS S3, and Azure Blob storage. We have also learned how to ingest data from the storage to a Spark dataframe, transform it using Spark methods or using SQL scripts, and then persist the data into tables. Finally, we have seen how to schedule our jobs with Azure Databricks jobs, and how to create a pipeline in ADF and trigger it using a defined variable.

In the next chapter, we will learn about Azure Delta Lake and how to use it to create reliable ETL pipelines using Azure Databricks.

4
Delta Lake with Azure Databricks

In this chapter, we will learn how to make use of Azure Databricks by showing how easy it is to work with Delta tables, as well as how to process data and integrate different solutions in Azure Databricks. We will begin by introducing Delta Lake and how to ingest data with it using either using partner integrations, the COPY INTO command, and the Azure Databricks Auto Loader. Then, we'll show you how to process the data once it has been loaded, as well as how we can use advance features in order to process and optimize ETLs that rely on streams of data.

We will learn how to store and process data efficiently using Delta by covering the following topics:

- Introducing Delta Lake
- Ingesting data using Delta Lake
- Batching table reads and writes in queries
- Querying past states of a table
- Streaming table read and writes

Technical requirements

To work with Delta Engine in this chapter, you will need an Azure Databricks Subscription, as well as a running cluster with an attached notebook.

Introducing Delta Lake

Using a data lake has become the de facto solution for many data engineering tasks. This storage layer is composed of files that can be arranged in a historical way instead of tables in a data warehouse. This has the benefit of decoupling storage from computing, which is the great advantage of data lakes. They are much cheaper than a database. The data that's stored in the data lake has no primary and foreign keys, making it hard to extract the information stored on it. Therefore, data lakes are seen as a solution where we only append new data. When trying to query or delete records, we need to go through all the files in the data lake, which could be a very resource-intensive and slow task.

This leads to data lakes being hard to update, and they may have problems when we try to use them in cases where data needs to be frequently queried. This includes customer or transactional data, financial applications that require robust data handling, or when we want changes in the data warehouse to be reflected on the records.

The data lake is being seen as a change in the architecture's paradigm, rather than a new technology, where we need to work on massive amounts of raw data that's produced by having several input sources dropping files into the data lake, which then need to be ingested. These files could contain structured, semi-structured, or unstructured data, which, in turn, is processed in parallel by different jobs that work concurrently, given the parallel nature of Azure Databricks. This can make it hard for the files to keep their integrity.

To tackle these limitations, Azure Databricks has introduced Delta Lake, which is a combination of Delta tables and a Delta Engine. Delta Engine can be used for the staging and processing layers, where it supports data streams, provides batch reads and writes to large queries, and provides analytic dashboards.

Delta Lake is a storage layer that sits on top of the data lake, such as **Azure Data Lake Storage** (**ADLS**) Gen2, and seeks to overcome the data lake's common limitations, such as being unable to execute SQL queries on the data or managing the ingestion of incoming files. It has several advantages, which includes automatically ingesting streams of data, data schema handling, data versioning thanks to time traveling, and faster processing than a conventional relational data warehouse, along with the benefit of being able to query the data. Delta Lake seeks to provide reliability for the big data lakes by ensuring data integrity with **Atomicity**, **Consistency**, **Isolation**, and **Durability** (**ACID**) transactions, allowing you to read and write the same file or table. The way in which Delta Lake works is very straightforward – you just need to change the way you store your files and specify Delta as the format of a table, which is a Spark proprietary file type that only works on the cloud.

We will deep dive into the specifics of these methods in the upcoming sections.

One thing to bear in mind is that Delta Lake can only be accessed from the Azure Databricks runtime, so we should not consider it as a replacement for a data warehouse.

Delta Lake provides several advantages related to how we work with the data stored in the lake. The most prominent features that make Delta Lake a great choice are as follows:

- **ACID transactions**: As we mentioned previously, isolation levels derived from ACID transactions allow different users to work concurrently with the same data, without them seeing inconsistencies. This is thanks to the use of transaction logs, which are included in the checkpoint support.

- Delta Lake provides scalable processing for metadata thanks to Azure Databricks' distributed computing nature.

- **Streaming and batch unification**: A table in Delta Lake is a batch table, as well as a streaming source and sink, making it a solution for a Lambda architecture. However, it also goes one step further since both batch and real-time data land in the same sink. Streaming data ingest, batch historic backfill, and interactive queries all just work out of the box.

- **Automatically enforce schema**: Automatically handles schema variations to prevent bad records being inserted during ingestion.

- **Time travel**: One of the most important features of Delta Lake is being able to access snapshots of data either as versions of a table or a specific timestamp. This allows us to fix any errors derived from possible bad data manipulation, handle auditing more easily, and provide a clear view of the changes that have been made to the data.

- **Upserts and deletes**: It supports DDL manipulations such as merges, delete, and updates thanks to the use of Spark APIs, enabling you to work with cases such as streaming data or SCD operations.

- Performance is boosted thanks to automatic data optimization, such as data skipping to read only relevant portions of data, file compression, catching frequently accessed data, and Z-Ordering.

In this section, we introduced Delta Lake and briefly mentioned what lead to its development, as well the tools that we can use to take full advantage of its main features. In the next section, we will dive into how we can start using Delta Lake by going through the methods used to load data into it from several sources.

Ingesting data using Delta Lake

Data can be ingested into Delta Lake in several ways. Azure Databricks offers several integrations with Partners, which provide data sources that are loaded as Delta tables. We can copy a file directly into a table, use AutoLoader, or create a new streaming table. Let's take a deeper look at this.

Partner integrations

Azure Databricks allows us to connect to different partners that provide data sources. These are easy to implement and provide scalable ingestion.

We can view the options that we have for ingesting data from **Partner Integrations** when creating a new table in the UI, as shown in the following screenshot:

Figure 4.1 – Ingesting data from Partner Integrations

Some of these integrations, such as Qlink, allow you to get data from multiple data sources such as Oracle, Microsoft SQL Server, and SAP and load them into Delta Lake. Of course, these integrations require you to have a subscription to these services and in most cases, the documentation will be provided by the supplier.

The COPY INTO SQL command

If we would like to copy data from a file into a delta table, we can use the COPY INTO command, which is an option that we can use to load files into a directory. Every time we run the command, it will skip files that have been already loaded. Let's take a look:

- As an example, if we want to load data into an already created table, we can do the following:

```
COPY INTO target_table
  FROM 'source_path'
```

- Otherwise, we can refer to the path of the table by passing the prefix delta to the target path:

```
COPY INTO delta.`target_path`
  FROM 'source_path'
```

The COPY INTO command also has several options in that we can specify a specific column from a file location, format options, and copy options.

The format of the source files to load should be either CSV, JSON, AVRO, ORC, or PARQUET. FILES refers to a list of filenames to load and they can have a length up to 1000. It cannot be specified with PATTERN, which is a regex pattern that identifies the files to load from the source directory.

FORMAT_OPTIONS and COPY_OPTIONS are options that can be passed to Apache Spark which for the latter are only 'force', which, if enabled, they load the files at hand, regardless of whether they were loaded previously:

- We can see the use of this option in the following command. Here, we are inserting data into a table, selecting the key, index, textData, and constant_value columns from the source path, specifying CSV as the format, specifying the pattern of the files required, and adding a formatting option so that we can add headers:

```
COPY INTO delta.`target_table_path`
  FROM (SELECT key, index, textData, 'constant_value'
FROM 'source_path')
  FILEFORMAT = CSV
  PATTERN = 'data_folder/file_[a-g].csv'
  FORMAT_OPTIONS('header' = 'true')
```

The COPY INTO command is commonly used in SQL and it's also a supported way to ingest data into a Delta Lake table.

In the next section, we will discuss how to automatically ingest data using Auto Loader, which is a Delta Lake feature that automatically ingests data into a Delta table.

Auto Loader

The problem of how to ingest data arises very frequently when working with data lakes. How the files are going to be read poses a problem if they are coming from different sources such as streams of data or IoT devices, which could lead to us needing to implement another service such as Event Hub and, from there, store it in the data lake. Another workaround would be to read the files from the data lake in batches or partitioned by time, which would increase complexity and hurt performance.

Auto Loader is an Azure Databricks feature that helps solve these issues by loading files in an incremental way, as soon as they arrive in storage. It does this using a structured source named cloudFiles. When specifying a path to work in, this can be set to process files that already exist on that directory. It can also detect files arriving by directory listing or file notification.

Auto Loader is a feature of Azure Databricks that allows us to process files as soon as they arrive in storage. This is done using a streaming source called **cloudFiles**, which reads files that arrive in a directory. We can also process existing files in that directory. We have two ways to detect files that are arriving in the storage directory: listing the files in the directory and using a file notification system. In both modes, Auto Loader will keep track of which files have been processed.

The file notification option detects files that are arriving, based on notifications that come from an Azure Event Grid subscription, to the file events in the storage account. Take into account that when you select this option, all the required services, such as storage account selected and Azure Event Grid, are created automatically and need to be cleaned manually later.

In both modes, files are processed exactly once, even if the file gets appended or overwritten. If a file is appended to or overwritten, it won't be clear which version of the file is being processed. Therefore, it is best to only use Auto Loader to ingest immutable files. Let's take a look at this:

- In the following Python command, we're reading a stream and using the useNotifications option, while set to true, to activate file notifications:

```
df = spark.readStream.format("cloudFiles") \
    .option("cloudFiles.useNotifications","true") \
    .schema(<schema>) \
    .load(<input-path>)
```

- In a more general way, we can read a stream into a DataFrame using the following Python command:

```
df = spark.readStream.format("cloudFiles").
option(<cloudFiles-option>, <option-value>) \
    .schema(<file_schema>) .load(<input-path>)
```

- Then, once we have read the stream, we can write it to a delta table by specifying a checkpoint location:

```
df.writeStream.format("delta").
option("checkpointLocation", <checkpoint-path>).
start(<output-path>)
```

When a new stream is created, a checkpoint is created in that directory. If the checkpoint is changed when the service is restarted, it will be assumed that a new stream has been created. The file notification process is more suitable when we're dealing with large amounts of data. This is because files are processed just once when they arrive, with no need for us to list the entire directory, as well as because multiple streaming queries can run in one input directory simultaneously.

One of the drawbacks of file notification is the need to spin up more resources in order to detect the files arriving in storage. It's more suitable when we're managing a large number of files, and it's also a more scalable solution.

Directory listing is the default mode of Auto Loader. This mode does what it says and lists the directory to identify files that arrive in it. It doesn't need to configure new permissions and is suitable for when a small number of files are arriving regularly in storage. However, if the number of files increases rapidly, it can quickly become difficult to list the entire directory.

If you need to change to file notification mode, you will need to restart the streaming process.

In the next section, we will learn how we can optimize DDL operations that are being performed on the data that we just fetched by batching table operations.

Batching table read and writes

When performing DDL operations such as merge and update on several large tables stored in databases with high concurrency, the transaction log can become blocked and lead to real outages in the data warehouse. All SQL statements are atomic, which means that modifications that take a long time will cause data to be locked for as long as the process is being executed, which can be a problem for real-time databases. To reduce the computational burden of these operations, we can optimize some of them so that they can run on smaller, easier-to-handle batches that only lock resources for brief periods.

Let's see how we can implement batch reads and writes in Delta Lake, thanks to the options provided by the Apache Spark API.

Creating a table

We can create Delta Lake tables either by using the Apache Spark `DataFrameWriter` or by using DDL commands such as `CREATE TABLE`. Let's take a look:

- Delta Lake tables are created in the metastore, which is also where the path of the table is saved, even though this is not the ground truth of the table, which still lies in Delta Lake:

```
CREATE TABLE data (
  input_date DATE,
  record_id STRING,
  record_type STRING,
  record_data STRING)
USING DELTA
```

- We can also perform this operation using Python. The following command creates a table named "data" in the metastore:

```
df.write.format("delta").saveAsTable("data")
```

- We can also create this table by identifying its path:

```
df.write.format("delta").save("/mnt/delta/data")
```

Here, we can see that Delta tables can be created either by common SQL commands or by simply using the PySpark API, which is convenient for reducing the changes required in the code so that it can adapt to using Delta tables.

The next section will show the many ways in which we can read the tables once they have been created.

Reading a Delta table

Delta tables can be loaded in two different ways:

- First, we can specify a table in the metastore as we would do in any SQL environment; that is, by using SELECT:

  ```
  SELECT * FROM data
  ```

  Here, data is the table that we want to read from.

- The other way we can load a table is to specify the path of the Delta file. We can do this by using delta a as prefix to the table's location:

  ```
  SELECT * FROM delta.'/mnt/delta_tables/data'
  ```

  Here, '/mnt/delta_tables/data' is the location of the table.

- As usual, we can also use Python to do this. To query a table by its name, we can run the following command:

  ```
  spark.table("data")
  ```

  Here, data is the table that we want to read from.

- Also, we can query the same table by referring to its path:

  ```
  spark.read.format("delta").load("/mnt/delta_tables/data")
  ```

  Here, '/mnt/delta_tables/data' is the location of the table.

One thing to bear in mind is that every time we read from a table, we are looking at the most recent snapshot of a table, without the need to use the REFRESH command. You can also specify partition statements to read data in a more efficient manner, as we will see in the following sections.

Partitioning data to speed up queries

You can speed up queries and DML operations by partitioning columns during processes. We can also apply this to Delta tables by using the PARTITIONED BY command on the column that we want to use as partition, which in many cases is a date column. If we specify LOCATION as a path to the Databricks filesystem, this creates the table in the metastore. If the table is not in the metastore, dropping the table won't remove the file specified in the location. Let's take a look:

- Here, we're creating a table in the Azure Databricks metastore and partitioning it by the input_date column:

```
CREATE TABLE data (
    input_date DATE,
    record_id STRING,
    record_type STRING,
    record_data STRING)
PARTITIONED BY (date)
LOCATION '/mnt/delta_files/data'
```

Again, we can also use Python for this operation.

- The following command performs the same operation, creating a table in the metastore:

```
df.write.format("delta").partitionBy("date").
saveAsTable("data")
```

- We can also perform the same operation by specifying the table's path:

```
df.write.format("delta").partitionBy("date").save("/mnt/
delta_files/data")
```

If tables are created in locations that already host a Delta table path, this can lead to the following scenarios. If only the table name and location are specified, it will be understood as if we just imported the data, and the created table will inherit the schema, partitioning, and properties of the table that was already in that path:

- The following example shows what would happen if `'/mnt/delta/data'` already hosted a table:

```
CREATE TABLE data
USING DELTA
LOCATION '/mnt/delta/data'
```

This also shows how the Delta Lake's automatic schema inference works.

If we had specified a schema and a partitioning, the Delta table would have checked that the input data matches the existing data. If it doesn't, an exception will be raised.

Querying past states of a table

As we mentioned previously, querying the past states of a table is a feature of Delta Lakes that is very useful when we're working with data that has been manipulated incorrectly, recovering deleted records, auditing, and in regulated industries. It also allows us to write queries that use different states in time, thanks to snapshot isolation.

This section describes how we can use these methods to query older versions of tables.

Using time travel to query tables

To query the previous state of a table, we can do the following:

- We can use the `AS OF` command, using SQL syntax:

```
SELECT * FROM table_name TIMESTAMP AS OF timestamp_
expression
```

The `table_name` parameter is the table's name. We can also pass any expression that can be cast as a timestamp or a date string to get the exact snapshot of data that we require.

- Another way to query the past states of a table is by using the `VERSION AS OF` command. To do this, we need to pass a specific version, which can be obtained from the output of `DESCRIBE HISTORY`. A full example of both methods is as follows:

```
SELECT * FROM table_name VERSION AS OF table_version
```

Here, `table_version` is a value that can be obtained by using `DESCRIBE HISTORY` on the table that we wish to query.

- If we would like to use this feature in a Python block of code in an Azure Databricks notebook, we can set the `DataFrameReader` option to one specific point in time or version of a table. To do this, we need to pass the `timestampAsOf` option:

```
df = spark.read.format("delta").option("timestampAsOf",
timestamp_string).load("/mnt/delta/data")
```

Here, `'/mnt/delta/data'` is the location of the table.

The same applies to accessing a specific version. We can do this by passing `versionAsOf` to the `DataFrameReader` option:

```
df = spark.read.format("delta").option("versionAsOf",
version).load("/mnt/delta/data")
```

It's easy to use time travel to select data from an earlier version of a table. Users can view the history of a table and see what the data looked like at that point by using either a version history number (as shown in the following code, when selecting the `VERSION AS OF 0` table data), or by timestamp:

- We can learn how to obtain the available versions and see the history of a table by reading the `HISTORY` property of the data using `DESCRIBE`:

```
DESCRIBE HISTORY data
```

- If we would like to access the snapshot of the table to see what the data looked like at a certain point in time, we can use `TIMESTAMP` combined with `AS OF`:

```
SELECT * FROM loan_by_state_delta TIMESTAMP AS OF '2020-
07-14 16:30:00'
```

Delta Lake guarantees that while querying the previous state of a table, all the transformations can be tracked, reproduced, and remain in a reversible state. This is especially useful while working with data that's bound by GDPR or CCPA regulations, in which it is required to have proof of deletion. We can also use this method to visualize every DML operation, such as updates and merges that have been made on this table.

Time travel is a safeguard against possible human mistakes or data mishandling and provides easy ways to debug errors, thanks to its transaction log. This allows us to go back in time, find the mechanism that created the error, action on that issue, and revert the data to an accurate state.

If we take into account that Delta tables automatically update themselves, it's good practice to set them to the latest version while executing jobs in Azure Databricks. Otherwise, the whole process may produce different results if the data gets updated during the execution process.

- We can execute a SQL query to get the latest version by using the following Python command:

```
latest_version = spark.sql("SELECT max(version).\
FROM (DESCRIBE HISTORY delta.`/mnt/delta/events`)").
collect()
df = spark.read.format("delta").option("versionAsOf",
latest_version[0][0]).load("/mnt/delta/events")
```

Another way to access a specific point in history or version of a Delta table is by using the @ syntax. This is a way to point to a certain state of a table during SQL and Python executions.

- An example of this is the following SQL command, in which we're using a timestamp to select all the rows of a table at a point in time:

```
SELECT * FROM events@20200101000000000
```

- We can also specify a version in the same way by adding v as a prefix to the version:

```
SELECT * FROM events@v114
```

The same applies to Python, where the @ syntax is passed directly to the path of the table we want to load.

- The following is an example of how to access a particular timestamp:

```
spark.read.format("delta").load("/mnt/delta/
events@20200101000000000")
```

- The same method applies to versions:

```
spark.read.format("delta").load("/mnt/delta/events@v114")
```

At this point, you might be wondering how far back in time you can go. The default is 30 days in the past, as long you haven't run the VACUUM command on the Delta table you want to access; otherwise, you won't be able to go further than 7 days into the past.

- This behavior can be configured by setting the delta.logRetentionDuration and delta.deletedFileRetentionDuration properties. For example, we can set delta.deletedFileRetentionDuration to 30 days by running the following command:

```
delta.deletedFileRetentionDuration = "interval 30 days".
```

This way, we can easily establish a retention policy for the available history of any delta table.

Making use of time travel allow us to have a failsafe in case we have performed the wrong operation on the data, and we want to recover the previous state. In the next section, we will learn how to leverage this feature while storing and processing data in Delta Lake.

Working with past and present data

Being able to access the previous time states of a table is especially useful when we want to recover information that has been removed or deleted, or when we want to monitor the growth of data in time. Let's take a look:

- The following is an example of this functionality. Here, we're inserting records that were previously deleted by passing the desired timestamp and merge for all the user IDs that match. These records are all from 7 days ago:

```
MERGE INTO target_table trg
  USING source_table TIMESTAMP AS OF date_sub(current_
date(), 7) src
  ON src.user_id = trg.user_id
  WHEN MATCHED THEN UPDATE SET *
```

- We can also use time travel to rapidly query the increase in the number of users in the last 30 days:

```
SELECT count(distinct user_id) - (
  SELECT count(distinct user_id)
  FROM targe_table TIMESTAMP AS OF date_sub(current_
date(), 30))
```

- Another example would be to override a particular subset of data. First, we must load the specific version that we want to use:

```
df = spark.read.format("delta").option("versionAsOf",
version).load("/mnt/delta/data")
```

- Then, we can select a particular date range, which in our case is the first 15 days of June to be overridden by the ones in the previous version:

```
df.write.format("delta").mode("overwrite")\
    .option("replaceWhere", "insert_date >= '2020-06-15'
AND insert_date <= '2020-06-15'")\
.save("/mnt/delta/data")
```

You can also state why the changes in the data have been made, which you can do by storing them as metadata in the commits that are produced when these changes take place. This metadata can be later accessed in the operations history.

- In the following example, we're replacing the complete version and state that we have made these changes to because the data was faulty:

```
df.write.format("delta").mode("overwrite") \
    .option("userMetadata", "updating-faulty-data").save("/
mnt/delta/data")
```

The "overwrite" mode will replace the delta table stored in "/mnt/delta/data". It will specify user metadata with the userMetadata option to describe why the change was made, in the same way as a Git commit message.

- The same can be done in SQL by setting the spark.databricks.delta.commitInfo.userMetadata option to the commit string:

```
SET spark.databricks.delta.commitInfo.
userMetadata=updating-faulty-data
INSERT OVERWRITE data SELECT * FROM correctData
```

In this section, we've learned how to use historical data to replace faulty data and roll back any changes that have been wrongfully made to Delta tables.

In the next section, we will learn how to automatically run validations and adapt to the schema of the data that's used in the Delta tables.

Schema validation

Changes in the schema of data can cause lots of problems, but they are also inevitable in cases where information has to be adapted to fit a new set of business rules, when tables need to be updated, and when improvements need to be introduced. Thankfully, Delta Lake has an adaptive schema evolution that works in combination with schema enforcement to fit and adapt to changes in the schema.

We can explicitly set this option to act by using the `mergeSchema` option when writing to a table. In the following example, we're doing this with the data DataFrame, which will be saved to Delta Lake:

```
data.write.option("mergeSchema","true"). \
 format("delta").mode("append").save(data_lake_path)
```

This option will cause every column that is present in the data but not in the specified Delta table to be added to the existing table. This is a good option when we don't want our ETL to break in production when we know that changes will be made in the future.

Delta Lake automatically checks that the schema of the data we are trying to insert into a table matches the schema of that table. This is done by following a set of rules, which are as follows:

- All of the columns in the DataFrame we are trying to insert exist in the table and have the same data types; otherwise, this will raise an exception.

- If we are using the `partitionBy` option while appending the data, it will be compared with the existing data and if any differences arise, it will exit with an error. Otherwise, if `partitionBy` is not being used, the data will be appended.

- The columns that don't exist in the data we are trying to insert are filled with null values.

- Column naming is case sensitive.

As we mentioned previously, if you want to update a table, you can do this by using the `overwriteSchema` option. This option allows you to change a column data type or drop a column, which, in turn, will require you to rewrite the table.

In the following example, we're changing the type of a column by using the overwriteSchema option in Python:

```
spark.read.table("/mnt/delta/data").withColumn("insertion_
date", col("insertion_date").cast("date")).write \
  .format("delta") .mode("overwrite").option("overwriteSchema",
"true").saveAsTable("/mnt/delta/data")
```

Here, we are using the Spark API to read a table located at /mnt/delta/data and change the data schema of the insertion_date column to a date data type. We're then saving the result back in the same table by specifying the overwriteSchema option.

In the next section, we will learn how to create a Delta table that automatically populates itself with streams of data from different sources.

Streaming table read and writes

Although we have already mentioned that structured streaming is a core part of the Apache Spark API, in this section, we will dive into how we can use it as a reliable stream processing engine, in which computation can be performed in the same way as batch computation is performed on static data. Along with Auto Loader, it will automatically handle data being streamed into Delta tables without the common inconveniences such as merging small files produced by low latency ingestion, running concurrent batch jobs when working with several streams, and, as we discussed earlier, keeping track of the files available for being streamed into tables.

Let's learn how to stream data into Delta tables in Azure Databricks.

Streaming from Delta tables

You can use a Delta table as a stream source for streaming queries. The query will then process any existing data in the table and any incoming data from the stream. Let's take a look:

- We can do this by directly using readStream and specifying the table path, as shown here:

  ```
  spark.readStream.format("delta").load("/mnt/delta/data")
  ```

- Alternatively, we can set readStream to read a specific data table:

  ```
  spark.readStream.format("delta").table("data")
  ```

Aspects such as the maximum size of micro batches provided in each stream or the amount of data to be computed in each batch can be controlled with options such as `maxFilesPerTrigger` and `maxBytesPerTrigger`.

Managing table updates and deletes

As we mentioned previously, if a table that is being used as a streaming source is updated, structured streaming will throw an exception. We can solve this issue by changing the `ignoreChanges` option, which will process updated tables so that the streaming won't be interrupted if any deletions or updates occur on the streaming table. We can also use the `ignoreDeletes` option if we only want to ignore deletions on the source data. Finally, if none of these options work, we can always directly delete the outputs and checkpoints and restart the stream.

An example of this situation would be when we're modifying a record in one of our tables using the `UPDATE` command. Modifying the record will modify the file that contains that table. If the `ignoreChanges` option is used, the new record will be propagated, along with any other records contained in the same file that haven't been modified.

Specifying an initial position

Delta Lake allows you to specify a certain point in the streaming source, without the need to process all the data again. The functionality of time travel allows us to specify either a `startingVersion` or a specific `startingTimestamp`, which will be our starting point for reading our stream. These options take effect on new streaming queries; otherwise, they are ignored.

The available versions can be obtained using the `DESCRIBE HISTORY` command. An example of this is accessing version 3 of our streaming data:

```
straming_events.readStream
   .format("delta")
   .option("startingVersion", "3")
   .load("/mnt/straming_events/user_events")
```

The `startingTimestamp` command specifies the point from which you will read onward in time. The input value can be a timestamp, a string that's been cast in timestamp format, a date string, or expressions that can be cast to a timestamp or date.

As we did previously, the following example specifies a starting timestamp to read our data from:

```
straming_events.readStream
  .format("delta")
  .option("startingTimestamp", "2020-06-15")
  .load("/mnt/straming_events/user_events")
```

It is worth noting that the schema of a table is always the latest schema, so there must be no incompatibilities when you're working with time traveling options.

Streaming modes

Delta Lake allows two different modes of streaming data to be used; that is, Append mode and Complete mode. The first is the default option, which means that new records are appended to the table. Let's take a look:

- This can be also set as an option by using `outputMode`, as shown here:

```
streaming_events.writeStream
  .format("delta").outputMode("append")
  .option("checkpointLocation", "/delta/streaming_
events/_checkpoints/etl-from-json")
  .start("/delta/straming_events/")
```

- Alternatively, we can use the table's name:

```
streaming_events.writeStream
  .format("delta").outputMode("append")
  .option("checkpointLocation", "/delta/streaming_
events/_checkpoints/etl-from-json")
  .table("streaming_events")
```

Complete mode means that the table is completely replaced in each new batch of data.

- We can also set this mode with `outputMode`, as shown here:

```
streaming_events.writeStream
  .format("delta").outputMode("complete")
  .option("checkpointLocation", "/delta/streaming_
events/_checkpoints/etl-from-json")
  .start("/delta/straming_events/")
```

If the latency requirements are low, you can set the process to work with one-time triggers instead of a streaming source. These triggers can be really helpful when we're work on things in a periodic way rather than in a constant stream. An example of this is when we're working with reports that only need to be updated when new data is available.

Optimization with Delta Lake

In this section, we will continue using the COVID dataset provided by Azure Databricks. Let's get started:

1. Our first task will be to load our data. To do this, we need to use `fs`, which will query the dataset:

   ```
   %fs
   ls dbfs:/databricks-datasets/COVID/coronavirusdataset/
   ```

   This should show all the available example datasets.

2. Then, we must select the `PatientInfo` file and load it into a `spark` DataFrame:

   ```
   file_path = "dbfs:/databricks-datasets/COVID/
   coronavirusdataset/PatientInfo.csv"
   df = spark.read.format("csv").load(file_path,header =
   "true",inferSchema = "true")
   display(df)
   ```

   Here, we can see the data that has been loaded into the DataFrame. This will be used to create the Delta table to be optimized:

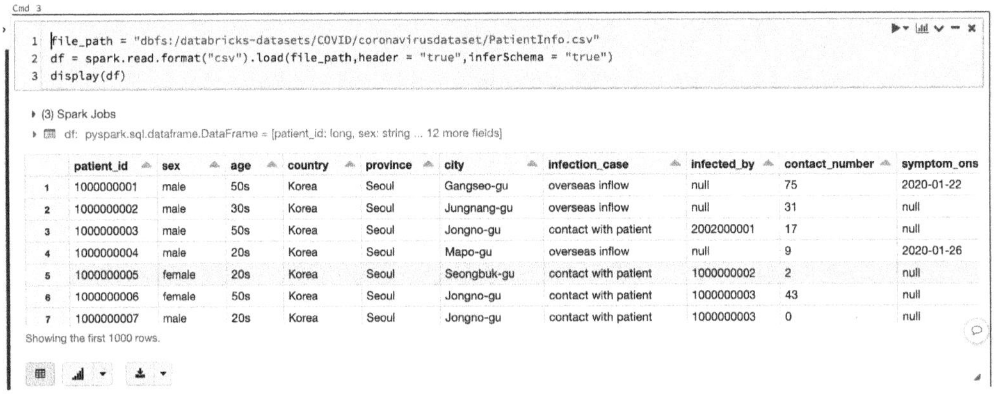

Figure 4.2 – The example patient DataFrame that we will be stored as a table

3. As usual, we can use the describe method to show the underlying structure of the DataFrame, which in this case has been inferred by Apache Spark:

```
df.describe().show()
```

The output of this command will show us the statistical summaries of the data that's been loaded:

Cmd 7

```
1  df.describe().show()
```

▸ (2) Spark Jobs

```
+-------+--------------------+----------+--------+----------+----------+------------------+------------------+
|summary|          patient_id|      date|province|      city|      type|          latitude|         longitude|
+-------+--------------------+----------+--------+----------+----------+------------------+------------------+
|  count|               10410|     10410|   10410|     10410|     10410|             10410|             10410|
|   mean| 2.08783917002805E9|      null|    null|      null|      null| 36.95588747070162|127.43422552353813|
| stddev|1.784859634692918E9|      null|    null|      null|      null|0.8408331877629417|0.7984479622723647|
|    min|          1000000001|2020-01-20|   Busan| Andong-si|   academy|          33.45464|           126.301|
|    max|          6100000133|2020-06-30|   Ulsan|Yuseong-gu|university|          38.19317|          129.4757|
+-------+--------------------+----------+--------+----------+----------+------------------+------------------+
```

Figure 4.3 – DataFrame information schema

4. Once the file has been loaded, we can write it to a Parquet file, which we will use to query later:

```
df.write.format("parquet").mode("overwrite").
partitionBy("province").save("/tmp/covid_parquet")
```

We're doing this so that we can write the table in a format that has been optimized to work in a data lake, such as Parquet.

5. Once the file has been written, we will read it again and use it to create a table. In order to read the DataFrame, we must run the following command:

```
covid_parquet = spark.read.format("parquet").load("/tmp/
covid_parquet")
```

Here, the table is being loaded from the /tmp/covid_parquet location.

6. Once we have loaded the Parquet file into Apache Spark, we can run a query on our data. We will run two queries. The will be based on a count and aggregation based on a date. To do this, we will use pyspark sql functions:

```
from pyspark.sql.functions import count
display(covid_parquet.groupBy("date").agg(count("*").
alias("TotalCount")).orderBy("TotalCount",
ascending=False).limit(20))
```

The preceding code shows how we can display the results of a simple count aggregation on the `patient_id` column by ordering by date. The following screenshot shows us the output of this operation:

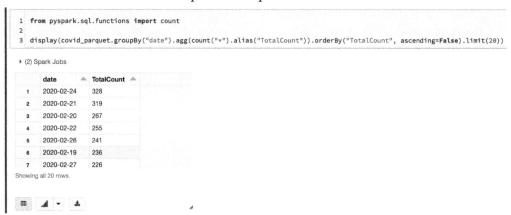

Figure 4.4 – Count aggregation on the date column

7. Under the same principle, we can also count the number of cases per province by using the aggregation on the province column:

Figure 4.5 – Aggregation over provinces

8. Then, we can query our data. A quick count on the DataFrame gives us 10,410 records:

```
Cmd 11

1  covid_parquet.count()

▶ (2) Spark Jobs
Out[77]: 10410
```

Figure 4.6 – Row count

9. Finally, we can write our data in delta file format to create our delta table. We can do this by specifying a partition by province, which is a column that we believe will be used frequently to filter data, as shown here:

```
covid_parquet.write.format("delta").mode("overwrite").
partitionBy("province").save("/delta/covid_delta/")
```

Cmd 12

```
1  covid_parquet.write.format("delta").mode("overwrite").partitionBy("province").save("/delta/covid_delta/")
```

▸ (5) Spark Jobs

Figure 4.7 – Writing the DataFrame in delta file format

10. The next step is to read the files in delta format to create our delta table in the metastore. We can read and display the data using the following command:

```
covid_delta = spark.read.format("delta").load("/delta/
covid_delta/")
display(covid_delta)
```

This way, we can see the results that have been loaded, as shown here:

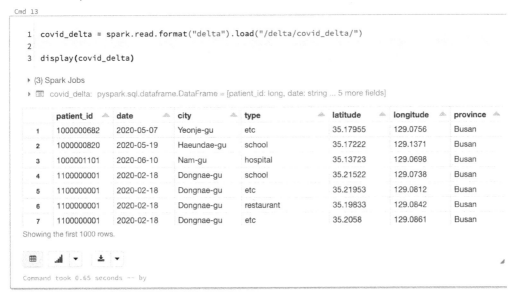

Cmd 13

```
1  covid_delta = spark.read.format("delta").load("/delta/covid_delta/")
2
3  display(covid_delta)
```

▸ (3) Spark Jobs

▸ 🖿 covid_delta: pyspark.sql.dataframe.DataFrame = [patient_id: long, date: string ... 5 more fields]

	patient_id	date	city	type	latitude	longitude	province
1	1000000682	2020-05-07	Yeonje-gu	etc	35.17955	129.0756	Busan
2	1000000820	2020-05-19	Haeundae-gu	school	35.17222	129.1371	Busan
3	1000001101	2020-06-10	Nam-gu	hospital	35.13723	129.0698	Busan
4	1100000001	2020-02-18	Dongnae-gu	school	35.21522	129.0738	Busan
5	1100000001	2020-02-18	Dongnae-gu	etc	35.21953	129.0812	Busan
6	1100000001	2020-02-18	Dongnae-gu	restaurant	35.19833	129.0842	Busan
7	1100000001	2020-02-18	Dongnae-gu	etc	35.2058	129.0861	Busan

Showing the first 1000 rows.

Command took 0.65 seconds -- by

Figure 4.8 – Reading the delta file into a Spark DataFrame

Now that our DataFrame is associated with the delta file, we can use it to create a delta table and take advantage of the optimization features of Delta Lake.

11. Next, we will use the DataFrame to create the delta table. First, we must run the following command, which drops the table if it already exists:

```
display(spark.sql("DROP TABLE IF EXISTS covid_delta"))
```

12. Then, we must create the table by specifying the location of the delta files with USING DELTA LOCATION:

```
display(spark.sql("CREATE TABLE covid_delta USING DELTA
LOCATION '/delta/covid_delta/'"))
```

Finally, we must apply ZORDER as an optimization technique. This is available in Delta Lake and seeks to reorder files so that they're as close as possible. In this case, our table is small, so running this operation won't have a significant effect. However, it is very useful when we're working with big tables with data scattered in different files.

1. We can run the optimization process by running OPTIMIZE on our delta table, along with the ZORDER BY statement. This statement is only applied to data columns, and not partition columns:

```
display(spark.sql("OPTIMIZE covid_delta ZORDER BY
(date)"))
```

The following screenshot shows the output of the entire optimize operation, in which we created and optimized a table by reading from a Delta table:

Figure 4.9 – Executing SQL queries on the delta table

2. Now, we can run queries while knowing that the layout of the data has been optimized, according to the filters that we will use to run queries. We also know that the number of files has been optimized to minimize the read of files and improve data skipping techniques:

Figure 4.10 – Count aggregation per day

3. Another way to create the table from the file is to use SQL commands. We can run the following commands in an Azure Databricks notebook cell in order to drop any existing table and create the table. We can do this by selecting all the columns and partitions by province:

```
%sql
DROP TABLE IF EXISTS covid_delta;
CREATE TABLE covid_delta
USING delta
PARTITIONED BY (province)
SELECT *
FROM delta.`/delta/covid_delta/`;
```

We should see **OK** as if this was successful, as shown in the following screenshot:

```
Cmd 19

1  %sql
2
3  DROP TABLE IF EXISTS covid_delta;
4  CREATE TABLE covid_delta
5  USING delta
6  PARTITIONED BY (province)
7  SELECT *
8  FROM delta.`/delta/covid_delta/`;


▶ (5) Spark Jobs
OK

Command took 13.02 seconds -- by
```

Figure 4.11 – Creating a delta table from the delta files using SQL

4. As we mentioned in the previous sections, we can also run SQL queries on the delta table that we have just created:

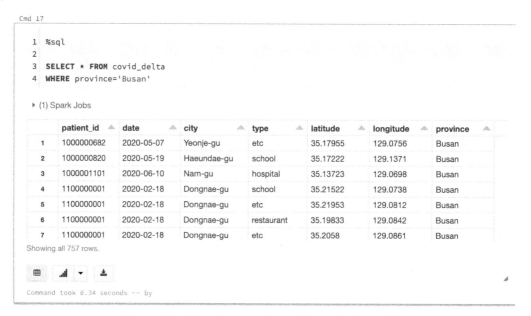

Figure 4.12 – Executing queries on SQL on the delta table

5. The path to our delta table can also be referenced using the file's location:

    ```
    SELECT * FROM delta.`/delta/covid_delta/`
    ```

    The SELECT statement works in the same way as it would do in a common SQL database:

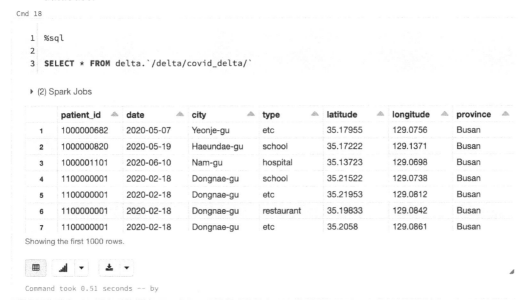

Figure 4.13 – Specifying a delta table by path

Delta tables have a number of optimization features that are enabled by default and others that can be triggered manually. This allows us to perform operations such as file size and location optimization, data skipping, and more. It is also easy to use, thanks to us being able to efficiently query files in our data lake.

Summary

Implementing a data lake is a paradigm change within an organization. Delta Lake provides a solution for this when we are dealing with streams of data from different sources, when the schema of the data might change over time, and when we need to have a system that is reliable against data mishandling and easy to audit.

Delta Lake fills the gap between the functionality of a data warehouse and the benefits of a data lake while also overcoming most of its challenges.

Schema validation ensures that our ETL pipelines maintain reliability against changes in the tables. It informs us of this by raising an exception if any mismatches arise and the data becomes contaminated. If the change was intentional, we can use schema evolution.

Time travel allows us to access historic versions of data, thanks to its ordered transaction log. This keeps track of every operation that's performed in Delta tables. This is useful when we need to define pipelines that need to query different points in time in history, revert changes, or track the origins of data mismatches.

In the next chapter, we will deepen our knowledge of data lakes by learning how we can leverage the features of Delta Engine, the computing engine in Delta Lake, to optimize our processes even further.

5
Introducing Delta Engine

Delta Engine is the query engine of Delta Lake, which is included by default in Azure Databricks. It is built in a way that allows us to optimize the processing of data in our Delta Lake in a variety of ways, thanks to optimized layouts and improved data indexing. These optimization operations include the use of **dynamic file pruning** (**DFP**), Z-Ordering, Auto Compaction, ad hoc processing, and more. The added benefit of these optimization operations is that several of these operations take place in an automatic manner, just by using Delta Lake. You will be using Delta Engine optimization in many ways.

In this chapter, you will learn how to make use of Delta Lake to optimize your Delta Lake ETL in Azure Databricks. Here are the topics on which we will center our discussion:

- Optimizing file management with Delta Engine
- Optimizing queries using DFP
- Using Bloom filters
- Optimizing join performance

Delta Engine is all about optimization, and file management can be one of our major drawbacks. In this chapter, you will learn how to use Delta Engine's features to improve the performance of your ETLs.

Technical requirements

To be able to work with Delta Engine in this chapter, you will need an Azure Databricks subscription.

Optimizing file management with Delta Engine

Delta Engine allows improved management of files in Delta Lake, yielding better query speed, thanks to optimization in the layout of the stored data. Delta Lake does this by using two types of algorithms: **bin-packing** and **Z-Ordering**. The first algorithm is useful when merging small files into larger ones and is more efficient in handling the larger ones. The second algorithm is borrowed from mathematical analysis and is applied to the underlying structure of the data to map multiple dimensions into one dimension while preserving the locality of the data points.

In this section, we will learn how these algorithms work, see how to implement them using commands that act on our data, and how to handle snapshots of data thanks to the time travel feature.

It is good to remember that although there are automatic optimizations that take place when we use Delta Lake, most of these optimizations do not occur automatically, and some of them must be manually applied. We will see how—and how often—to do this.

Merging small files using bin-packing

We have discussed in the previous chapters difficulties that may arise from working with several small files that arrive at our data lake from different sources. Delta Engine provides us a workaround for this issue by allowing us to coalesce small files into larger files.

Bin-packing is, like Z-Ordering, an algorithm borrowed from mathematical analysis that was originally conceived to optimize how to pack objects of different volumes into a finite number of bins while using the smallest number of bins possible. We can already think of how this can be applied to the problem of packing smaller files into the smallest number of larger, more efficient files. They are as follows:

- We can apply this algorithm to a Delta table by using the OPTIMIZE command on a specific table path, as shown in the following code example:

```
OPTIMIZE delta.`/data/data_events`
```

Running this operation in a table will show as output the path of the table affected by the operation, along with metrics of the number of files that were modified during the optimization, as illustrated in the following screenshot:

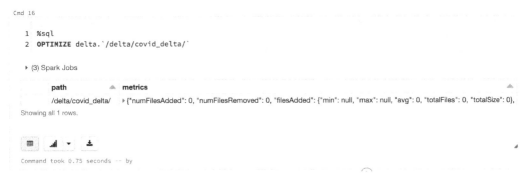

Figure 5.1 – Specifying a Delta table to optimize by path

- As we are used to doing, we can also just pass the table name, like this:

```
OPTIMIZE data_events
```

This table is small, so the optimization will not actually change anything on the files holding the table data, as can be seen in the following screenshot:

Figure 5.2 – Specifying a Delta table to optimize by name

- If we have many files but we want to focus on a certain number of them, we can optimize a portion of the data by explicitly passing a WHERE clause. This will cause the OPTIMIZE command to just act on a subset of data, as illustrated in the following code snippet:

```
OPTIMIZE data_events WHERE insert_date >= '2020-06-15'
```

When the OPTIMIZE command is run, we will see afterward a series of file statistics related to the reorganization of files that has occurred, such as removed files and added files, and even some Z-Ordering statistics.

Delta Lake will optimize the aforementioned tables, yielding better query speed, thanks to the fact that we will be scanning a smaller number of files to read our table. One thing to bear in mind is that the bin-packing optimization, triggered by the OPTIMIZE command, is an idempotent action, which is a fancy word for saying that the execution of this command will take effect only the first time that it is triggered on a dataset.

When this action takes place on a table, readers of the table that is being optimized will not be affected, thanks to the **Snapshot Isolation** feature of Delta tables. The integrity of the table is ensured even when the underlying file structure is reordered by the algorithm and the transaction log is modified accordingly, so queries will yield the same result before and after the OPTIMIZE command has been run on a table. Any stream of data related to a table that is being optimized will not be affected by this command either.

The goal of this algorithm is to produce a series of evenly sized files that provide a balance between the size and number of them in the storage.

So, if this reorganization is so efficient and doesn't affect the data stored in a table, why isn't OPTIMIZE automatic? The answer to this question lies in the fact that to coalesce several files into a single larger one, you need to have upfront those small files in the directory, otherwise there won't be anything to coalesce; therefore, OPTIMIZE can't be run automatically. Another answer is that compression is also an expensive operation in terms of computational, storage, and time resources, so this might affect the ability of Azure Databricks to execute low-latency streams when a table that is being optimized is the source of a stream.

A final point of discussion relates to how often Delta tables should be optimized. This, of course, is a trade-off between performance and the cost of the resource that is used to run the optimization. If your priority is performance, this operation should take place more often or at regular intervals such as daily intervals; otherwise, you can prioritize cost and run this less often.

Skipping data

As new data arrives, Delta Lake will keep track of file-level statistics related to the **input/output (I/O)** granularity of the data. These statistics are obtained automatically, storing information on minimum and maximum values at query time. The point of this is to identify columns that have a certain level of correlation, plan queries accordingly, and avoid unnecessary I/O operations. During lookup queries, the Azure Databricks Data Skipping feature will leverage the obtained statistics to skip files that are not necessary to read in this operation.

The granularity can be tuned and it integrates very well with partitioning, even though these are independent operations.

Bear in mind that Data Skipping is an indexing probabilistic technique that, as with Bloom filters (which will be discussed later in the chapter), can lead to false positives.

There is no need to configure Data Skipping because it is activated by default and applied whenever possible, meaning also that the performance of this operation is bound to the underlying structure of your data. What we can configure is how many columns we want to track during the statistics—which by default are the first 32, a value that can be modified using the `dataSkippingNumIndexedCols` option. Consider that keeping track of columns is a resource-intensive operation, so adding more columns or tracking columns with long string data types will have an impact on the overhead when writing files.

A workaround for these issues would be to change the type of a long string column using `ALTER TABLE` or move it beyond the threshold established by the `dataSkippingNumIndexedCols` option, to be skipped when collecting statistics.

Thanks to Data Skipping, we can reduce the amount of data needed to be read in each query, yielding an increase in the performance of our ETL pipelines and reducing time and costs.

In the next section, we will learn about Z-Ordering and how it can be leveraged to improve performance even more.

Using Z-order clustering

Sometimes, data can be scattered across files, which is not practical when we run queries against it. This leads to an increase in the size of metadata, as well as problems listing directories, and file compression issues.

To handle this problem, we can use partitioning techniques (as we will see in the next section) that attempt to make the distribution of our information more even. In turn, the performance of these techniques depends on how the information is scattered across files, so before running them you are advised to apply Z-Ordering.

Z-Ordering is a feature of Delta Lake that seeks to allocate related data in the same location. This can improve the performance of our queries on the table in which we run this command because Azure Databricks can leverage it with data-skipping methods to read and partition files more efficiently. We can use Z-Ordering in Azure Databricks using the ZORDER BY command:

- In the following code example, we will optimize the data table on the event_type column:

```
OPTIMIZE data
ZORDER BY (event_type)
```

By running this operation in a **Structured Query Language (SQL)** cell in a notebook, we should get as a result the modifications made to optimize the table according to the specified partition. From the following screenshot example, we can see that our table was small and we didn't really need to optimize the underlying file structure of the table:

Figure 5.3 – Applying ZORDER on a Delta table

- We can also reference Delta tables by path, as illustrated in the following code snippet:

```
OPTIMIZE delta.'path_to_data'
ZORDER BY (event_type)
```

By running the preceding code in a SQL cell in an Azure Databricks notebook, we should see the resulting metrics of the operation in the same way as before when the table was specified by name, as illustrated in the following screenshot:

Figure 5.4 – Applying ZORDER on a Delta table, specifying it by path

Z-Ordering can also be applied to a subset of data by passing a condition.

- In the following code example, we will only optimize data in the `event_type` column between today and yesterday:

```
OPTIMIZE data
WHERE insert_date >= current_timestamp() - INTERVAL 1 day
ZORDER BY (event_type)
```

In the following screenshot, we can see that by running this operation in a SQL cell in a notebook, we will see as a result the path of the table and the metrics of the operation in the same way as in the previous examples:

Figure 5.5 – Applying ZORDER to a specific portion of data

Columns that have many unique values, as well as columns used to filter by very frequently, are the most recommended columns to optimize by running ZORDER. We can also specify several columns by passing the column names, separated as parameters, to the ZORDER command.

Running ZORDER multiple times will lead to multiple executions, but this is considered an incremental operation. Running ZORDER after new data has arrived won't affect data that was already optimized.

One thing to bear in mind is that running Z-Ordering does not relate to data size. The goal of the operation is to relocate related files but not to evenly distribute data as this can also lead to data skewness, an issue that we will see how to solve in the next sections.

Managing data recency

As mentioned in the previous chapter, when querying a Delta table we are reading the latest state of that table unless specified otherwise, thanks to the fact that Delta tables are auto updated. This is a process that is run automatically, but in case we want to, we can modify this behavior if we don't need it. This can happen when performing historical analysis, whereby we do not require tables to be previously updated. Doing this will yield lower latency when making queries, which in short means that queries will run faster.

This behavior can be set using the spark.databricks.delta.stalenessLimit option, which takes as a parameter a time string value and applies this to the current session we are working on, so it won't affect other users reading that table. The time parameters are expressed in **milliseconds (ms)** and can range, for example, from 1 hour to 1 day. If this staleness limit is surpassed, the table state will be updated.

It's good to mention that the parameter doesn't block a table from being updated—it only avoids a query to wait for a table to be updated.

Understanding checkpoints

We have already discussed checkpoints when talking about streaming sources on tables, but we haven't completely talked about how they work.

Delta Lake uses checkpoints to aggregate the state of a Delta table, to make this a starting point for a computation when we require the latest state of the table. Checkpoints are written by default every 10 commits and they are stored in the _delta_log directory in the path of a table.

This mechanism saves a lot of time because otherwise, when querying a table, to get the latest state Delta Lake would have to read large amounts of **JavaScript Object Notation** (**JSON**) files that represent the transaction log of a table to obtain the last state.

In a nutshell, checkpoints hold information relating to all the transactions and modifications made to a table until the current state, checking that previous actions that have been canceled out by more recent ones get removed from the checkpoint. These kinds of actions are called **invalid actions** and they are removed using a policy known as **rules for reconciliation**. This is done to reduce the overall size of the log and to make it more efficient to be read when reconstructing snapshots. As we can see, checkpoints are crucial for snapshot isolation and accessing previous versions of the data.

Checkpoints are also the place where statistical information on column usage is stored to be used during—for example—Data Skipping.

When looking at a log directory, we will see that it has a naming convention, as illustrated in the following code snippet:

```
00000000000000001025.checkpoint.parquet
```

We can see that the name of the checkpoint is composed by a prefix that refers to the version of the table that it holds and the file format of the table. In a more general manner, we could say that the name of the checkpoint of the *n* version of a table that has `table_ format` would be `n.checkpoint.table_format`.

Given this format and the nature of the transaction log, it would be quite difficult to list all the checkpoints in the directory (which can be in their thousands) if a user would like to access the latest checkpoint. Therefore, Delta Lake stores the last checkpoint in a `_ delta_log/_last_checkpoint` file to provide easy access for constructing the latest snapshot of a table.

As mentioned before, when working with structured streaming, the stream can restart in the case of a failure if checkpoints are enabled as shown:

- To do this, we can set the `delta.checkpoint.writeStatsAsStruct` option to `true` in our streaming table, as follows:

```
ALTER TABLE our_table SET TBLPROPERTIES
('delta.checkpoint.writeStatsAsStruct' = 'true')
```

We can run this in a SQL cell in a notebook, which should return `OK` if the operation was successful, as illustrated in the following example:

```
Cmd 17

1  %sql
2  ALTER TABLE covid_delta SET TBLPROPERTIES
3  ('delta.checkpoint.writeStatsAsStruct' = 'true')
4

▶ (3) Spark Jobs
OK

Command took 1.76 seconds -- by
```

Figure 5.6 – Setting checkpoint settings on a Delta table

- Or, when working with Delta tables, we can use the path of the table with a `delta` prefix, as illustrated in the following code snippet:

```
ALTER TABLE delta.`path_to_our_table` SET TBLPROPERTIES
('delta.checkpoint.writeStatsAsStruct' = 'true')
```

- We can also use the Apache Spark `checkpointLocation` option when working with a streaming dataframe by running the following Python command, which writes a streaming dataframe named `data` in Parquet format and sets the checkpoint location path. Each query must have a unique checkpoint:

```
data.writeStream.format("parquet")
  .option("path", output_path)
  .option("checkpointLocation", checkpoint_path)
  .start()
```

Having checkpoints in our Structured Streaming allows us to have a backup in case of possible failures. If a stream gets restarted, it will automatically continue from where it was left.

Automatically optimizing files with Delta Engine

Auto Optimize is a feature of Delta Engine to further optimize Delta tables. It automatically compacts small files during **data manipulation language** (**DML**) actions in big tables, consolidating files and keeping metadata clean. We have two options to apply Auto Optimize: **Optimized Writes** and **Auto Compaction**. Optimized Writes tries to arrange table partitions evenly into 128 **megabytes** (**MB**)-sized files representing each partition to improve writes. Auto Compaction is an OPTIMIZE command with the size of compaction set to 128 MB, which is run after each write operation.

Auto Optimize keeps frequently modified tables optimized, which is very useful when dealing with—for example—low-latency streams of data or frequent table merges.

We need to enable Auto Optimize by enabling the options in our_delta_table.autoOptimize.optimizeWrite and our_delta_table.autoOptimize.autoCompact on specific tables:

- To enable this option on existing tables, we can use the TBLPROPERTIES SQL command to set those properties to true, as in the following example:

```
ALTER TABLE our_table SET TBLPROPERTIES (delta.
autoOptimize.optimizeWrite = true, delta.autoOptimize.
autoCompact = true)
```

We can run this command in a SQL-specified cell in the notebook, as illustrated in the following screenshot:

```
Cmd 18

1  %sql
2  ALTER TABLE covid_delta
3  SET TBLPROPERTIES
4  (delta.autoOptimize.optimizeWrite = true, delta.autoOptimize.autoCompact = true)

▶ (3) Spark Jobs
OK
Command took 1.60 seconds -- by
```

Figure 5.7 – Enabling Auto Optimize on a Delta table

- We can also define these options as true by default on all new Delta tables, using the following SQL configuration:

```
set spark.databricks.delta.properties.defaults.
autoOptimize.optimizeWrite = true.

set spark.databricks.delta.properties.defaults.
autoOptimize.autoCompact = true.
```

- If we would like to apply this only to our working session, we can use the following code to set the following Spark configurations:

```
spark.databricks.delta.optimizeWrite.enabled
spark.databricks.delta.autoCompact.enabled
```

One reason to have to enable these options manually is that if we have many concurrent DML operations on the same table from different users, this may cause conflicts in the transaction log, which can be dangerous because this will not cause any failure or retry.

Another thing to keep in mind is that on big tables, having Auto Optimize enabled on a table can be used in combination with `OPTIMIZE` and does not replace the running of `ZORDER`.

Let's see in more detail how to use Optimized Writes and Auto Compaction, as follows:

- **Optimized Writes**: When Optimized Writes is enabled on a table, Delta Engine will try to reorder the table into approximately 128 MB files for each table partition, which can vary according to the data schema, to balance the number of files being written in parallel. This process reorders the files based on the partition structure of the table. This option is useful when having data streams with a latency of minutes.

- **Auto Compaction**: As mentioned before, Auto Compaction is run after each write, so if it is enabled on a table it is not necessary to call `OPTIMIZE` after each write operation. It is a good idea to enable this option if we do not have a periodically scheduled `OPTIMIZE` job running on the working table.

Most of the time when optimizing data processing in Azure Databricks, it is easy to just focus on the transformations. But we should always remember that Databricks uses distributed computing, having the need to distribute data across workers. Making sure that data is neatly organized and its storage has been optimized will lead to better performance metrics, not only for storage but also for processing of the data.

Using caching to improve performance

Caching is an operation in which we keep data stored closer to where it will be processed, to improve performance. It can be applied in two different ways in Azure Databricks: Delta caching and Apache Spark caching. Depending on the situation we are dealing with, we can leverage the specific characteristics of each option to improve the reading speed of our tables.

The good thing about this is that there is no need to make a choice on which option to use because they can be used simultaneously.

Let's dive into the characteristics of both options and look at how they are applied to improve performance.

Delta and Apache Spark caching

Delta and Apache Spark have different characteristics, which might be summarized as follows:

- **Delta caching**: The Delta cache automatically makes copies of remote Parquet files into the local node storage to accelerate its processing. The remote locations that it can access are **Databricks File System** (**DBFS**), **Hadoop Distributed File System** (**HDFS**), Azure Blob storage, Azure Data Lake Storage Gen1, and Azure Data Lake Storage Gen2. It can operate faster than the Spark cache thanks to optimized decompression and an output format consistent with processing requirements. Data can be preloaded using a CACHE statement. Data is then stored on the local disk and can be read quickly thanks to a **solid-state drive** (**SSD**).

- **Spark caching**: The Spark cache can store results from subqueries and handles several file formats besides Parquet. Queries and tables to be cached must be manually specified. Data is stored in the memory, which can hurt performance.

We can enable Delta caching by selecting one of the **Delta Cache Accelerated** workers from the cluster configuration, as illustrated in the following screenshot:

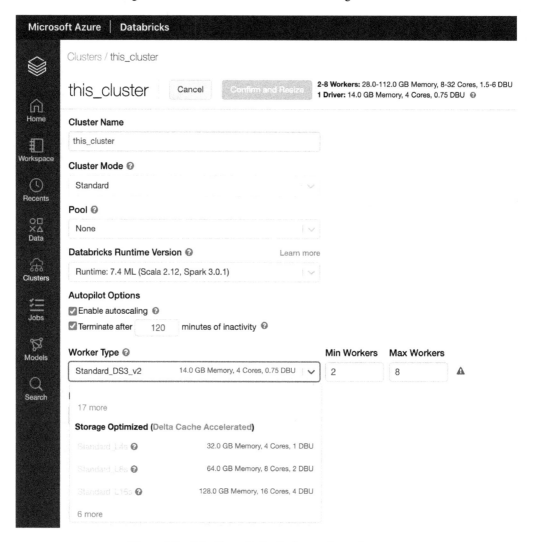

Figure 5.8 – Selecting a Delta Cache accelerated storage

Then, we continue as follows:

- To enable the Spark cache, we need to run the cache option on our dataframe—as shown in the following Python command—where we cache a dataframe called `data`:

```
data.cache()
```

- The cache will call the persist method, passing MEMORY_DISK as a parameter, meaning that this is the same as the following Python command:

```
data.persist(StorageLevel.MEMORY_AND_DISK)
```

We have several parameters to pass as storage levels, which are outlined as follows:

- DISK_ONLY: This will persist only serialized data on disk.

- MEMORY_ONLY: This will persist deserialized data in memory only, achieving a very good computational performance but having as a limitation the amount of memory available for storage.

- MEMORY_AND_DISK: This will persist data in memory and if not enough memory is available, it will evict data to be stored on disk. This is the default mode in Spark, where it will store as much data as possible in the floating memory, with the rest evicted to the disk. The computation time will be faster than just with a disk but slower than with memory only.

- OFF_HEAP: Data is persisted in off-heap memory, which is memory that is not supervised by the **Java virtual machine** (**JVM**) garbage collector. The data will be read faster but it should be used carefully because it's the user who must specify the allocated memory.

Having fine-grained control over where the cached data is stored in the worker allows us to also have fine-grained control over the performance of the algorithms making use of the data. Data that is stored in the local worker memory will be read faster than from the disk but will be able to hold smaller amounts of data. OFF_HEAP memory should be used carefully because its data is also used by the JVM for internal processes.

Caching a subset of the data

We can also explicitly select only a portion of data to be cached in the worker local storage by using the CACHE SELECT command in the following way:

```
CACHE SELECT col1, col2, col3
FROM our_table
WHERE our_condition
```

In the same way, we can specify a Delta table using a `delta` prefix and the table path, as illustrated in the following code snippet:

```
CACHE SELECT col1, col2, col3
FROM delta. 'our_table'
WHERE our_condition
```

In the following screenshot example, we run this in a SQL cell in a notebook, specifying the condition as a specific province of the COVID dataset:

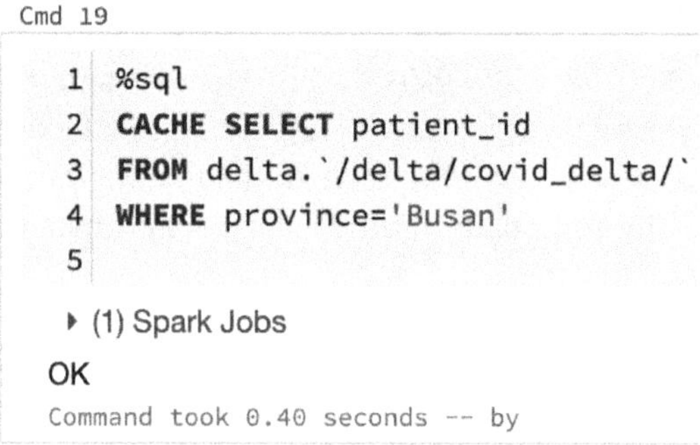

```
Cmd 19

1  %sql
2  CACHE SELECT patient_id
3  FROM delta.`/delta/covid_delta/`
4  WHERE province='Busan'
5

▶ (1) Spark Jobs

OK
Command took 0.40 seconds -- by
```

Figure 5.9 – Specifying a portion of a Delta table to be cached

This will only record from the specified columns that comply with our condition to be cached, thus preventing subsequent queries from scanning remote files.

Doing this is not necessary when working with Delta tables because data will be cached when queried, but it can be useful when you need to boost your query's performance.

A simple example of this would be to cache a table named data, as follows:

```
CACHE SELECT * FROM user_data
```

Or, we can also do this on portions of data that will be frequently used during the session, as follows:

```
CACHE SELECT user_id, user_cohort FROM user_data WHERE user_
cohort=10
```

Up next, we will see how we can configure the behavior of the Delta cache.

Configuring the Delta cache

It is recommended to always use Delta cache-accelerated clusters for worker instances, which are optimized instances running SSD disks. These instances are already configured to work with the Delta cache as default, but you can manually configure its parameters using Apache Spark options during the session creation:

- To establish a maximum disk usage, which is set by default to half of the worker local storage, you can use the `maxDiskUsage` option, measured in bytes, as shown in the following example, where we set it to 50g:

```
spark.databricks.io.cache.maxDiskUsage 50g
```

- We can also specify the maximum allocated space measured in bytes for metadata with the `maxMetaDataCache` option, as shown in the following example, where it is set to 1g:

```
spark.databricks.io.cache.maxMetaDataCache 1g
```

- We can also disable the compression, which is enabled by default, as follows:

```
spark.databricks.io.cache.compression.enabled false
```

- To disable the Delta cache, we can use the following Scala command:

```
spark.conf.set("spark.databricks.io.cache.enabled",
"false")
```

It is good to know that when disabling the Delta cache we will not lose data that was already stored in the worker local storage, but we will avoid more data being added.

In the next section, you will read about DFP, which is a Delta Lake feature that improves query performance.

Optimizing queries using DFP

DFP is a Delta Lake feature that automatically skips files that are not relevant to a query. It is a default option in Azure Databricks and works by collecting data about files in Delta Lake, without the need to explicitly state that a file should be skipped on a query, improving performance by making use of the granularity of the data.

The behavior of DFP concerning whether a process is enabled or not, the minimum size of a table, and the minimum number of files needed to trigger a process can be managed by the following options:

- `spark.databricks.optimizer.dynamicPartitionPruning` (default is `true`): Whether DFP is enabled or not.

- `spark.databricks.optimizer.deltaTableSizeThreshold` (default is `10 GB`): The minimum size of the Delta table that activates DFP.

- `spark.databricks.optimizer.deltaTableFilesThreshold` (default is `1000`): Represents the number of files of the Delta table on the probe side of the join required to trigger DFP. If the number of files is low, it's not worthwhile triggering the process.

DFP can be used for non-partitioned tables, nonselective joins, and when performing DML operations on them.

The improvement obtained by applying DFP is correlated with the level of granularity of the data. Therefore, it might be a good idea to apply Z-Ordering when working with DFP.

Using DFP

To first understand how we apply DFP, let's see an example of where we manually use literal statements and another Delta Lake feature called **dynamic partition pruning (DPP)**.

During the physical planning phase, Delta Lake creates an executable plan for the query. This plan distributes the computation across the clusters of many workers. This plan is optimized to only scan files relevant to the query.

If we were to run a query but we wanted to explicitly filter the results based on a condition. A more concrete example of this would be to select col1, col2, and col3 from our table, and apply a filtering condition on one of the columns based on a maximum value, as illustrated in the following code snippet:

```
SELECT col1,col2,col3
FROM our_table
WHERE col1> max_value
```

Delta Lake keeps information on the minimum and maximum value in our columns, so it will automatically skip files that are not relevant to our query, based on our filtering condition.

This is possible because the filtering contains a literal condition (which in our case is a maximum value condition), so Delta Lake can plan and optimize the file scanning. If the filtering condition is part of a join, Delta Lake cannot plan which files to skip because the filtering condition is the result of another query. This is called DPP and is a previous feature of Delta Lake that, like DFP, seeks to improve the performance of our queries.

The following kinds of queries, whereby we first join two or more tables, are very common when using data models based on a star schema:

```
SELECT col1, col2, col3
FROM fact_table src
JOIN dim_table trg
ON src.id_col=trg.id_col
WHERE col1=our_condition
```

In this case, the condition acts on the dimension table and not on the fact table. Delta Lake will have to scan all the files on the fact table, scan all the files in the dimension table, apply the filtering condition, and then perform the join.

In this case, DPP won't work, but if we have enabled DFP it will reorder the execution plan of the query to be able to check in advance the condition on both tables, which gives the possibility of skipping files on both tables and increasing the performance of the query.

DFP is enabled by default in Azure Databricks, and it is applied to queries based on the following rules:

- The join strategy is BROADCAST HASH JOIN.
- The join type is INNER or LEFT-SEMI.
- The inner table is a Delta table.
- The number of files in the inner table is big enough to justify applying DFP.

Depending on the selected options, DFP will be applied in a join operation.

Using Bloom filters

Bloom filters are a way of efficiently filtering records in a database based on a condition. They have a probabilistic nature and are used to test the membership of an element in a set. We can encounter false positives but not false negatives. These filters were developed as a mathematical construct, to be applied when the amount of data to scan is impractical to be read, and are based on hashing techniques.

Delta Lake provides us with the ability to apply Bloom filters on our queries to further improve performance. We will see how they work at a basic level and how they can be applied in Delta Lake.

Understanding Bloom filters

As mentioned in the introduction to this section, Bloom filters are probabilistic data structures used to test if an element belongs to a category or not. This structure is a fixed-length bit array that is populated using a hash function, which maps the information into ones and zeros. The length of the array depends on the number of false positives that are tolerated. Boom filters have constant complexity and require very little space relative to the items.

To map data into this probabilistic data structure, hashing functions that allow us to obtain a *fingerprint* of the data are used, meaning that this hash is (almost) unique and can be used to compare or identify data. These hashing functions are one-way operations and must always provide the same output for the same data, and—if possible—the output values must be uniformly and randomly distributed so that similar inputs won't give the same results.

These hashes can be later compared bitwise to check for similarity.

In the following screenshot example, we compare if the *W* element belongs to the *{x, y, z}* group. To do this, we will compare the positions that have one of the resulting hash array of hashing *W* against the combined positions filled by all the hashes of the group. If all the *1*s in the *W* hash are in the positions filled by the group's hash, then *W* must be in the set. If any of the bits in the *W* hash fall into a *0*, then they don't belong to the set:

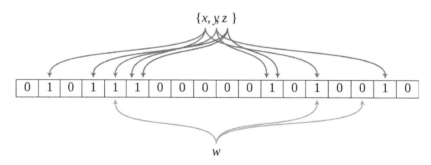

Figure 5.10 – Comparing hashes to check membership (image taken from https://en.wikipedia.org/wiki/Bloom_filter#/media/File:Bloom_filter.svg)

In this case, the *W* element isn't in the set since it hashes to a bit position containing *0*.

We can see that Bloom filters are a fast way of providing an answer if an element belongs to a set and that they are of great importance in data engineering. Next, we will see how we can implement Bloom filters in Delta Lake to ensure that our queries perform well.

Bloom filters in Azure Databricks

Azure Databricks Delta Lake uses Bloom filters to check if a file matches a particular filter when running queries to enable data skipping. It does this by creating Bloom filter index files associated with each file, with up to five levels, which can quickly tell Delta Lake the probability of the file containing the required data or if it doesn't match at all. These files are single-row Parquet files stored in a subdirectory in the path of the data file.

If a Bloom index file doesn't exist for a given file, this file will always be read. The size of the Bloom filter index files is, of course, proportional to the **false positive rate** (**FPR**) selected because a lower rate will always require more bits in the array to compare, which is—for example—5 bits for a given 10% FPR.

_delta_index contains the index files, which have a naming format that is the same as for the data with the added index.v1.parquet suffix.

We can enable or disable Bloom filters by setting the spark.databricks. io.skipping.bloomFilter.enabled session configuration to either true or false, as illustrated in the following screenshot. This option is already enabled by default in Azure Databricks:

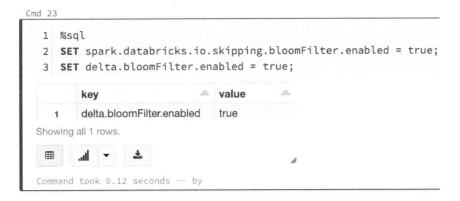

Figure 5.11 – Setting Bloom filters as default for new Delta tables

Creating a Bloom filter index

We can create Bloom filters using the CREATE BLOOMFILTER INDEX command. This applies to all columns on a table or only a subset of them.

The syntax to create a Bloom filter is shown in the following code snippet:

```
CREATE BLOOMFILTER INDEX ON TABLE our_table
FOR COLUMNS(col1 OPTIONS(fpp=0.1), col2 OPTIONS(fpp=0.2))]
```

In the case of the COVID dataset used in the previous chapters, we can create a Bloom filter in the patient_id column, as shown in the following screenshot:

```
Cmd 24

1  %sql
2  CREATE BLOOMFILTER INDEX
3  ON TABLE covid_delta
4  FOR COLUMNS(patient_id OPTIONS (fpp=0.1, numItems=50000000))

▶ (3) Spark Jobs
OK
Command took 1.97 seconds -- by
```

Figure 5.12 – Creating Bloom filters for a Delta table

We can also reference the same command using the path to a Delta table, as illustrated in the following code snippet:

```
CREATE BLOOMFILTER INDEX ON TABLE delta.'path_to_table'
FOR COLUMNS(col1 OPTIONS(fpp=0.1), col2 OPTIONS(fpp=0.2))]
```

In this case, we are creating Bloom filters for the col1 and col2 columns with FPRs of 10% and 20%, respectively.

We can also pass the numItems option, which is the number of distinct items a file can contain, and the maxExpectedFpp option, which controls the maximum FPR for which a Bloom filter is not written to disk. That said, we have created a way to more quickly query our data written to memory, which we must compute just once.

We can also drop all Bloom filters from a table using the DROP BLOOMFILTER INDEX command. If we would like to delete our recently created Bloom filter, we can run the following code:

```
DROP BLOOMFILTER INDEX ON TABLE our_table
FOR COLUMNS(col1 , col2)
```

This will remove all the Bloom filters in the col1 and col2 columns. Bear in mind that running VACUUM on a table will also remove all Bloom filters.

Optimizing join performance

Performing joins on tables can be a resource-expensive operation. To improve the performance of such operations, we can select a subset of the data or correct possible drawbacks, such as having a disproportionate distribution of file sizes in our data. Solving these issues can improve performance and lead to more efficient use of distributed computing power.

Azure Databricks Delta Lake allows optimization of join operations by providing range filtering and correcting skewness in the distribution of the file size of the data in our tables.

Range join optimization

Joins are used frequently, so optimizing these operations can lead to a great improvement in the performance of our queries. Range join optimization is the process of specifying that a join needs to be performed on a subset of data given by a range.

Range join optimization is applied when join operations have a filtering condition whose type is either a numeric or datetime type, and can be understood as an interval and have a bin-size tuning number.

An example of an interval can be seen in the following code snippet:

```
SELECT * FROM data_points JOIN ranges ON data_points.p BETWEEN
ranges.start_range and ranges.end_range;
```

To use range join optimization, we need to establish the required bin size. To do this, we use a hint or set this up in the Spark session configuration, as it will not run unless specified. The bin-size number will be interpreted in days when using a date type and in seconds when using a timestamp.

Enabling range join optimization

We enable range join optimization by explicitly passing a hint with the parameter corresponding to the bin size. The hint is called by using a specific RANGE_JOIN syntax on a SELECT statement and receives as parameters the target table and the bin size. Besides tables, we can also specify views and subqueries, taking into account that this might hurt performance.

In the next example, the range join is established on a table with a bin size of 20:

```
SELECT /*+ RANGE_JOIN(data_points, 20) */ *
FROM data_points JOIN ranges ON data_points.p >= ranges.start_
range AND points.p < ranges.end_range;
```

You can also specify in range joins the bin size as a percentile of the data. In the following case, this is 10%:

```
SELECT /*+ RANGE_JOIN(r1, 0.1) */ *
FROM (SELECT * FROM ranges WHERE ranges.amount < 100) r1,
ranges r2
WHERE r1.start_range < r2.start_range + 100 AND r2.start_range
< r1.start_range + 100;
```

Finally, we can also specify the range join in timestamps and datetime formats, as discussed before, using the BETWEEN command:

```
SELECT /*+ RANGE_JOIN(data_points_c, 100) */ *
FROM data_points_a
   JOIN data_points_b ON (data_points.point_id = data_points_b.
point_id)
   JOIN data_points_c ON (data_points_a.ts BETWEEN data_
points_c.start_time AND data_points_c.end_time)
```

How effective the optimization is will depend on having the right bin size. A bin size that is too small in relation to the intervals might lead to overlaps between intervals. We can choose appropriate bin sizes by looking at the percentiles found in the column that we will be using as a filter. This can be done using the APPROX_PERCENTILE command.

In the following code example, we are looking at the .25, .5, and .75 percentiles of data_ranges with intervals between the end and start column:

```
SELECT APPROX_PERCENTILE(CAST(end - start AS DOUBLE),
ARRAY(.25, .5, .75)) FROM data_ranges
```

If the length of our intervals is uniform, we can set the bin size to the most common interval length.

The bin size is a parameter that is highly dependent on the distribution of data, so the recommendation is to test different approaches to finding the right bin size to leverage all the parallel processing computational advantages.

Skew join optimization

We have a skewness problem when a partition of our data is much bigger than the rest. This can yield problems when we execute our process on that partition because it will take much more time to work on this partition. This is especially bad on distributed computation systems such as Azure Databricks because many tasks are run in parallel, and having one task taking longer than the rest can block executions that have a dependency on this task.

This uneven workload problem affects join operations mostly because big tables need to be reshuffled. It is possible that this problem will affect our queries if some executions take longer than others.

This issue can be tackled using skew join optimization hints in Azure Databricks in the same way as we have seen with range join optimization.

Hints are specified using a /* syntax to start and /* to end in a SELECT statement, passing the SKEW command, which has as a parameter the column that we want to optimize.

An example of this is shown in the following code snippet:

```
SELECT /*+ SKEW('orders') */ * FROM data, users WHERE user_id =
our_condition
```

Delta Lake uses the information in hints to optimize the execution plan for our query. We can also apply it to subqueries, as shown in the following code example, where we run a subquery to filter our data:

```
SELECT /*+ SKEW('C1') */ * FROM (SELECT * FROM data WHERE user_
id < 100) C1, orders
  WHERE C1.user_id = our_condition
```

We can also specify relationships and columns with the SKEW command. We will see how to do this in the next section.

Relationships and columns

Skew join optimization is a computationally intensive operation that must be run only when it is required. That is why the argument of SKEW is column names. When running joins, it is useful to then just run SKEW on the column that we are using to do this.

In the following code example, we apply the skew hint to a single column, which is user_id:

```
SELECT /*+ SKEW('orders', user_id) */ *
  FROM orders, users
  WHERE user_id = our_condition
```

We can also apply the hint to multiple columns, as follows:

```
SELECT /*+ SKEW('orders', ('user_id', 'o_region_id')) */ *
  FROM orders, users
  WHERE user_id = our_condition AND o_region_id= c_region_id
```

You can also specify skew values in the hint. Depending on the query and data, the skew values might be known (for example, because they never change) or might be easy to find out. Doing this reduces the overhead of skew join optimization. Otherwise, Delta Lake detects them automatically.

Values can be also specified on the hints, depending on the type of query. The more specific the parameters passed to SKEW, the less overhead there will be for the optimization task.

In the following code example, we run a query on a single column specifying a single skew value:

```
SELECT /*+ SKEW('orders', 'user_id', 0) */ *
 FROM orders, users
  WHERE user_id = our_condition
```

We can also pass several skew values, as follows:

```
SELECT /*+ SKEW('orders', 'user_id', (0, 1, 2)) */ *
 FROM orders, users
  WHERE user_id = our_condition
```

And finally, we can also specify several columns and—respectively—several values to those columns, as follows:

```
SELECT /*+ SKEW('orders', ('user_id', o_region_id), ((0, 1001),
(1, 1002))) */ *
  FROM orders, users
    WHERE user_id = our_condition AND o_region_id= c_region_id
```

We can see that we can correct skewness in our data in a selective manner, applying this transformation only in columns that we select that also comply with the stated condition. That way, we can optimize in a more effective manner.

Summary

In this chapter, we have discussed how we can apply Delta Engine features such as Z-Ordering, Data Skipping, and data caching, among others, to improve the layout structure of our data, leading to an overall improvement in query performance. We can leverage these tools to reorganize files into lesser, much more compact ones, in which data is distributed according to how frequently it is accessed. We can also create Bloom filters on data columns to improve the speed in which we run our queries.

In the next chapter, we will deepen our knowledge of how to ingest streams of data in our Delta tables.

6
Introducing Structured Streaming

Many organizations have a need to consume large amounts of data continuously in their everyday processes. Therefore, in order to be able to extract insights and use the data, we need to be able to process this information as it arrives, resulting in a need for continuous data ingestion processes. These continuous applications create a need to overcome challenges such as creating a reliable process that ensures the correctness of the data, despite possible failures such as traffic spikes, data not arriving in time, upstream failures, and so on, which are common when working with continuously incoming data or transforming data without consistent file formats that have different structure levels or need to be aggregated before being used.

The most traditional way of dealing with these issues was to work with batches of data executed in periodic tasks, which processed raw streams and data and stored them into more efficient formats to allow queries on the data, a process that might incur high latency.

In this chapter, we will learn about Structured Streaming, which is the Apache Spark **application programming interface** (**API**) that allows us to make computations on streaming data in the same way you would express a batch computation on static data. Structured Streaming is a fault-tolerant engine that allows us to perform optimized operations for low-latency **Structured Query Language** (**SQL**)-like queries on the data thanks to the use of Spark SQL, while ensuring data consistency.

We can use Structured Streaming when creating ETL pipelines that will allow us to filter, transform, and clean data as it comes in a raw structure, formatting data types into more efficient storage formats, applying efficient partition strategies based on relevant columns, and so on. Specifically, in this chapter, we will look at the following topics:

- Using the Structured Streaming API
- Using different sources available in Azure Databricks when dealing with continuous streams of data
- Recovering from query failures
- Optimizing streaming queries
- Triggering streaming query executions
- Visualizing data on streaming dataframes
- Example on Structured Streaming

Without further ado, we will start our discussion by talking about Structured Streaming models and how can we leverage their architecture to improve performance when working with streams of data.

Technical requirements

This chapter will require you to have an Azure Databricks subscription available to work on the examples, as well as a notebook attached to a running cluster.

Let's start by looking into Structured Streaming models in more detail to find out which alternatives are available to work with streams of data in Azure Databricks.

Structured Streaming model

A Structured Streaming model is based on a simple but powerful premise: any query executed on the data will yield the same results as a batch job at a given time. This model ensures consistency and reliability by processing data as it arrives in the data lake, within the engine, and when working with external systems.

As seen before in the previous chapters, to use Structured Streaming we just need to use Spark dataframes and the API, stating which are the I/O locations.

Structured Streaming models work by treating all new data that arrives as a new row appended to an unbound table, thereby giving us the opportunity to run batch jobs on data as if all the input were being retained, without having to do so. We then can query the streaming data as a static table and output the result to a data sink.

Structured Streaming is able to do this, thanks to a feature called **Incrementalization**, which plans a streaming execution every time that we run a query. This also helps to figure out which states need to remain updated every time new data arrives.

We define when the tables will be updated using specific triggers when using Structured Streaming dataframes. Each time a trigger is activated, Azure Databricks will update the table incrementally by creating new rows for each file that arrives in our data source.

Finally, the last part to be defined are the output modes. These modes allow us to determine how to write incrementally to external systems such as **Simple Storage Service (S3)** or a database in three different ways, outlined as follows:

- **Append mode**: Only new rows are appended. This mode doesn't allow changes to existing rows.

- **Complete mode**: The entire table is rewritten to the external storage every time there is an update.

- **Update mode**: Only rows that have changed since the last trigger are written to the external storage, which allows in-place updates such as SQL tables.

Structured Streaming output tables provide consistent results based on a sequential stream, and failure tolerance is also ensured when working with output sinks. Using the Complete output mode, for example, out-of-order data will be handled by updating the records in a SQL table. We can establish a threshold to avoid overly old data updating a table.

Structured Streaming is a fault-tolerant system that allows queries to be run incrementally on streaming data, as explained in the following examples:

- For example, if we want to read **JavaScript Object Notation** (**JSON**) files that arrive continuously at an S3 location, we write the following code:

```
input_df = spark.readStream.json("s3://datalogs")
```

- We can operate on the Spark dataframe and do time-based aggregations to finally write to a SQL database, as follows:

```
input_df.groupBy($"message", window("update_time",
                                "1 hour")).count()
        .writeStream.format("jdbc").start(jdbc_conn_
string)
```

- If we run this as a batch operation, the only changes will be about how we write and read the data. For example, to read the data just once, run the following code:

```
input_df = spark.read.json("s3://datalogs")
```

- Then, we can do operations using the standard Spark DataFrame API and write to a SQL database, as illustrated in the following code snippet:

```
inputDF.groupBy("message", window("update_time",
                                "1 hour")).count()
        .writeStream.format("jdbc").save(jdbc_conn_string)
```

In the next section, we will see in detail how to use the Structured Streaming API in more detail, including how to perform interactive queries and windowed aggregations, and we will discuss mapping and filtering operations.

Using the Structured Streaming API

Structured Streaming is integrated into the PySpark API and embedded in the Spark DataFrame API. It provides ease of use when working with streaming data and, in most cases, it requires very small changes to migrate from a computation on static data to a streaming computation. It provides features to perform windowed aggregation and for setting the parameters of the execution model.

As we have discussed in previous chapters, in Azure Databricks, streams of data are represented as Spark dataframes. We can verify that the data frame is a stream of data by checking that the isStreaming property of the data frame is set as true. In order to operate with Structured Streaming, we can summarize the steps as read, process, and write, as exemplified here:

1. We can read streams of data that are being dumped in, for example, an S3 bucket. The following example code shows how we can use the readStream method, specifying that we are reading a **comma-separated values** (**CSV**) file stored in the S3 **Uniform Resource Interface** (**URI**) passed:

    ```
    input_df=spark.readStream.csv("s3://data_stream_uri")
    ```

    The resulting PySpark dataframe, input_df, will be the input table, continuously extended with new rows as new files arrive in the input directory. We will suppose that this table has two columns—update_time and message.

2. Now, you can use the usual data frame operations to transform the data as if it were static. For example, if we want to count action types of each hour, we can do this by grouping the data by message and 1-hour windows of time, as illustrated in the following code snippet:

    ```
    counts_df = input_df.groupBy($"message", window($"update_
    time ", "1 hour")).count()
    ```

    The new dataframe, counts_df, is our result table, which has message, window, and counts columns, and will be continuously updated when the query is started. Note that this operation will yield the resulting hourly counts even if input_df is a static table. This is a feature that provides a safe environment to test a business logic on static datasets and seamlessly migrate into streaming data without changing the logic.

3. Finally, we write this table to a sink and start the streaming computation by creating a connection to a SQL database using a **Java Database Connectivity** (**JDBC**) connection string, as illustrated in the following code snippet:

    ```
    final_query = counts_df.writeStream.format("jdbc").
    start("jdbc://...")
    ```

The returned query is a StreamingQuery query, which is a handle to the active streaming execution running in the background. It can later be used to manage and monitor the execution of our streaming query. Beyond these basics, there are many more operations that can be done in Structured Streaming.

Mapping, filtering, and running aggregations

Structured Streaming allows the use of mapping, filtering, selection, and other methods to transform data. In the previous example, we used one of these features: time-based aggregations on data through the Spark API. As we have seen in our previous example, aggregations such as windowing can be expressed as a simple group by operations on a data frame. The use of such operations will be exemplified in the next section, but it is good to remember that Structured Streaming allows us to perform mapping, filtering, and other data-wrangling methods in the same way as with a common PySpark dataframe.

In the next section, we will see some examples of how these operations are performed using Structured Streaming.

Windowed aggregations on event time

When working with data that has a temporal dimension, compute operations often need to be windowed to certain periods of time, which in cases such as sliding windows overlap with each other. For example, we might need to perform operations over a window of 1 hour that slides forward every 5 minutes.

Windows are specified using the `window` function in a PySpark dataframe. For example, change to a sliding window approach (shown in the example previously given in the Structured Streaming model section) by doing the following:

```
input_df.groupBy("message", window("update_time", "1 hour",
                                    "5 minutes")).count()
```

In the way our previous example was structured, the resulting information was formatted as (`hour, message, count`); now, it will have the form (`window, message, count`). Out-of-order data arriving late will be processed and the results will be updated accordingly, and if we are using an external data sink in Complete mode, the data will be updated there as well.

Merging streaming and static data

As we have mentioned before, working with Structured Streaming means working with Spark data frames, which is straightforward and allows us to use streaming and static data in combination. For example, if we have a table named users and we want to append it to a streaming data frame we can do the following, where we bring in a table called `users` and append it to our `input_df` dataframe:

```
users_df = spark.table("users")
input_df.join(users_df, "user_id").groupBy("user_name",
hour("time")).count()
```

We could also create this static data frame using a query and run it in the same batch and streaming operations.

Interactive queries

In Structured Streaming, we can make use of the results from computations directly to interactive queries using the Spark JDBC server. There is an option to use a small memory sink designed for small amounts of data, whereby we can write results to a SQL table and then query the data in it. For example, we can create a table named `message_counts` that can be used as a small data sink that we can later query.

We do this by creating an in-memory table, as follows:

```
message_counts.writeStream.format("memory")
    .queryName("user_counts")
    .outputMode("complete")
    .start()
```

And then, we use a SQL statement to query it, like this:

```
%sql
select sum(count) from message_counts where message='warning'"
```

In the next section, we will discuss all the different sources from where we can read stream data.

Using different sources with continous streams

Streams of data can come from a variety of sources. Structured Streaming provides support from extracting data from sources such as Delta tables, **publish/subscribe (pub/sub)** systems such as Azure Event Hubs, and more. We will review some of these sources in the next sections to learn how we can connect these streams of data into our jobs running in Azure Databricks.

Using a Delta table as a stream source

As mentioned in the previous chapter, you can use Structured Streaming with Delta Lake using the `readStream` and `writeStream` Spark methods, with a particular focus on overcoming issues related to handling and processing small files, managing batch jobs, and detecting new files efficiently.

When a Delta table is used as a data stream source, all the queries done on that table will process the information on that table as well as any data that has arrived since the stream started.

In the next example, we will load both the path and the tables into a dataframe, as follows:

```
df = spark.readStream.format("delta").load("/mnt/delta/data_
events")
```

Or, we can do the same by referencing the Delta table, as follows:

```
df = spark.readStream.format("delta").table('data_events')
```

One of the features of Structured Streaming in Delta Lake is that we can control the maximum number of new files to be considered every time a trigger is activated. We can control this by setting the `maxFilesPerTrigger` option to the desired number of files to be considered. Another option to set is the rate limit on how much data gets to be processed in every micro-batch. This is controlled by using the `maxBytesPerTrigger` option, which controls the number of bytes processed each time the trigger is activated. The number of bytes to be processed will approximately match the number specified in this option, but it can slightly surpass this limit.

One thing to mention is that Structured Streaming will fail if we attempt to append a column or modify the schema of a Delta table that is being used as the source. If changes need to be introduced, we can approach this in two different ways, outlined as follows:

- After the changes are made, we can delete the output and checkpoint and restart the stream from the start.

- We can use the `ignoreDeletes` option to ignore operations that causes the deletion of data, or `ignoreChanges` so that the stream doesn't get interrupted by deletions or updates in the source table.

For example, suppose that we have a table named `user_messages` from where we stream out data that contains date, `user_email`, and message columns partitioned by date. If we need to delete data at partition boundaries, which means deleting using a `WHERE` clause on the partition column, the files are already segmented by that column. Therefore, the delete operation will just remove those files from the metadata.

So, if you just want to delete data using a `WHERE` clause on a partition column, you can use the `ignoreDeletes` option to avoid having your stream disrupted, as illustrated in the following code snippet:

```
data_events.readStream
    .format("delta").option("ignoreDeletes", "true")
    .load("/mnt/delta/user_messages")
```

However, if you have to delete data based on the `user_email` column, then you will need to use the following code:

```
data_events.readStream
    .format("delta").option("ignoreChanges", "true")
    .load("/mnt/delta/user_messages")
```

If you update a `user_email` variable with an `UPDATE` statement, the file that contains that record will be rewritten. If the `ignoreChanges` option is set to true, this file will be processed again, inserting the new record and other already processed records into the same file, producing duplicates downstream. Thus, it is recommended to implement some logic to handle these incoming duplicate records.

In order to specify a starting point of the Delta Lake streaming source without processing the entire table, we can use the following options:

- `startingVersion`: The Delta Lake version to start from

- `startingTimestamp`: The timestamp to start from

We can choose between these two options but not use both at the same time. Also, the changes made by setting this option will only take place once a new stream has been started. If the stream was already running and a checkpoint was written, these options would be ignored:

- For example, suppose you have a `user_messages` table. If you want to read changes since version 10, you can use the following code:

```
data_events.readStream
  .format("delta").option("startingVersion", "10")
  .load("/mnt/delta/user_messages")
```

- If you want to read changes since 2020-10-15, you can use the following code:

```
data_events.readStream
  .format("delta")option("startingTimestamp", "2020-10-15")
  .load("/mnt/delta/user_messages")
```

Keep in mind that if you set the starting point using one of these options, the schema of the table will still be the latest one of the Delta table, so we must avoid creating a change that makes the schema incompatible. Otherwise, the streaming source will either yield incorrect results or fail when reading the data.

Using Structured Streaming allows us to write into Delta tables and, thanks to the transaction log, Delta Lake will guarantee exactly once processing is done, even with several queries being run against the streaming table.

By default, Structured Streaming runs in **Append** mode, which adds new records to the table. This option is controlled by the `outputMode` parameter.

For example, we can set the mode to `"append"` and reference the table using the `path` method, as follows:

```
data_events.writeStream.format("delta")
  .outputMode("append")
  .option("checkpointLocation", "/delta/data_events/_checkpoints/etl-from-json")
  .start("/delta/data_events")
```

Or, we could use the `table` method, as follows:

```
data_events.writeStream.format("delta")
   .outputMode("append")
   .option("checkpointLocation", "/delta/data_events/_
checkpoints/etl-from-json")
   .table("data_events")
```

We can use **Complete** mode in order to reprocess all the information available on a table. For example, if we want to aggregate the count of users by `user_id` value every time we update the table, we can set the mode to complete and the entire table will be reprocessed in every batch, as illustrated in the following code snippet:

```
spark.readStream.format("delta")
   .load("/mnt/delta/data_events")
   .groupBy("user_id").count()
   .writeStream.format("delta")
   .outputMode("complete")
   .option("checkpointLocation", "/mnt/delta/messages_by_user/_
checkpoints/streaming-agg")
   .start("/mnt/delta/messages_by_user")
```

This behavior can be better controlled using one-time triggers that allow us to have fine-grained control on what the actions are that should trigger the table update. We can use them in applications such as aggregation tables or with data that needs to be processed on a daily basis.

Azure Event Hubs

Azure Event Hubs is a telemetry service that collects, transforms, and stores events. We can use it to ingest data from different telemetry sources and to trigger processes in the cloud.

There are many ways to establish a connection. One way is to use `ConnectionStringBuilder` to make your connection string. To do this, follow these steps:

1. The following Scala code shows how we can establish a connection using `ConnectionStringBuilder` by defining parameters such as the event hub name, **shared access signature (SAS)** key, and key name:

```
import org.apache.spark.eventhubs.ConnectionStringBuilder
val connections_string = ConnectionStringBuilder()
    .setNamespaceName("your_namespace-name")
    .setEventHubName("event_hub-name")
    .setSasKeyName("your_key-name")
    .setSasKey("your_key build
```

    In order to create this connection, we need to install the Azure Event Hubs library in the Azure Databricks workspace in which you are using Maven coordinates. This connector is updated regularly, so to check the current version, go to **Latest Releases** in the *Azure Event Hubs Spark Connector* project README file.

2. After we have set up our connection, we can use it to start streaming data into the Spark dataframe. The following Scala code shows how we first define the Event Hubs configuration:

```
val event_hub_conf = EventHubsConf(connections_string)
    .setStartingPosition(EventPosition.fromEndOfStream)
```

3. Finally, we can start streaming into our dataframe, as follows:

```
var streaming_df =
  spark.readStream
    .format("eventhubs").options(event_hub_conf.toMap)
    .load()
```

4. You can also define this using Python and specifying the connection string to Azure Event Hubs and then start to read the stream of that directly, as follows:

```
conf = {}
conf["eventhubs.connectionString"] = ""
stream_df = spark
    .readStream
    .format("eventhubs")
```

```
        .options(**conf)
        .load()
```

Here, we have created a connection and used this to read the stream directly from the Azure event hub specified in the connection string.

Establishing a connection between Azure Event Hubs allows you to aggregate the telemetry of different sources being tracked into a single Spark data frame that can be later on be processed and consumed by an application or stored for analytics.

Auto Loader

Auto Loader is an Azure Databricks feature used to incrementally process data as it arrives at Azure Blob storage, Azure Data Lake Storage Gen1, or Azure Data Lake Storage Gen2. It provides a Structured Streaming source named `cloudFiles` that, when given an input directory path on the cloud file storage, automatically processes files as they arrive in it, with the added option of also processing existing files in that directory.

Auto Loader uses two modes for detecting files as they arrive at the storage point, outlined as follows:

- **Directory listing**: This lists all files into the input directory. It might become too slow when the number of files in the storage grows too large.

- **File notification**: This creates an Azure Event Hubs notification service to detect new files when they arrive in the input directory. This is a more scalable solution for large input directories.

The file detection mode can be changed when the stream is restarted. For example, if the **Directory listing** mode starts to slow down, we can change it to **File notification**. On both modes, Auto Loader will keep track of the files to guarantee that these are processed only once.

The `cloudFiles` option can be set in the same way as for other streaming sources, as follows:

```
df = spark.readStream.format("cloudFiles") \
    .option(<cloudFiles-option>, <option-value>) \
    .schema(<schema>).load(<input-path>)
```

After changing the file detection mode, we can start a new stream by writing it back again to a different checkpoint path, like this:

```
df.writeStream.format("delta") \
   .option("checkpointLocation", <checkpoint-path>).
start(<output-path>)
```

Next, we will see how we can use Apache Kafka in Structured Streaming as one of its sources of data.

Apache Kafka

Pub/sub messaging systems are used to provide asynchronous service-to-service communication. They are used in serverless and microservices architectures to construct event-driven architectures. In these systems, messages published are immediately received by all subscribers to the topic.

Apache Kafka is a distributed pub/sub messaging system that consumes live data streams and makes them available to downstream stakeholders in a parallel and fault-tolerant fashion. It's very useful when constructing reliable live-streaming data pipelines that need to work with data across different processing systems.

Data in Apache Kafka is organized into topics that are in turn split into partitions. Each partition is an ordered sequence of records that resembles a commit log. As new data arrives at each partition in Apache Kafka, each record is assigned a sequential ID number called an **offset**. The data in these topics is retained for a certain amount of time (called a **retention period**), and it's a configurable parameter.

The seemingly infinite nature of the system means that we need to decide the point in time from where we want to start read our data. For this, we have the following three choices:

- **Earliest**: We will start reading our data at the beginning of the stream, excluding data that is older than the retention period deleted from Kafka.

- **Latest**: We will process only new data that arrives after the query has started.

- **Per-partition assignment**: With this, we specify the precise offset to start from for every partition, to control the exact point in time from where the processing should start. This is useful when we want to pick up exactly where a process failed.

The startingOffsets option accepts just one of these options and is only used on queries that start from a new checkpoint. A query that is restarted from an existing checkpoint will always resume where it was left off, except when the data at that offset is older than the retention period.

One advantage when using Apache Kafka in Structured Streaming is that we can manage what to do when the stream first starts and what to do if the query is not able to pick up from where it left off because the data is older than the retention period, using the startingOffsets and failOnDataLoss options, respectively. This is set with the auto.offset.reset configuration option.

In Structured Streaming, you can express complex transformations such as one-time aggregation and output the results to a variety of systems. As we have seen, using Structured Streaming with Apache Kafka allows you to transform augmented streams of data read from Apache Kafka using the same APIs as when working with batch data and integrate data read from Kafka, along with data stored in other systems including such as S3 or Azure Blob storage.

In order to work with Apache Kafka, it is recommended that you store your certificates in Azure Blob storage or Azure Data Lake Storage Gen2, to be later accessed using a mount point. Once your paths are mounted and you have already stored your secrets, you can connect to Apache Kafka by running the following code:

```
streaming_kafka_df = spark.readStream.format("kafka")
  .option("kafka.bootstrap.servers", ...)
  .option("kafka.ssl.truststore.location",<dbfs-truststore-
location>)
  .option("kafka.ssl.keystore.location", <dbfs-keystore-
location>)
  .option("kafka.ssl.keystore.password", dbutils.secrets.
get(scope=<certificate-scope-name>,key=<keystore-password-key-
name>))
  .option("kafka.ssl.truststore.password", dbutils.secrets.
get(scope=<certificate-scope-name>,key=<truststore-password-
key-name>))
```

To write into Apache Kafka, we can use writeStream on any PySpark dataframe that contains a column named value and, optionally, a column named key. If a key column is not specified, then a null-valued key column will be automatically added. Bear in mind that a null-valued key column may lead to uneven partitions of data in Kafka, so you should be aware of this behavior.

We can specify the destination topic either as an option to `DataStreamWriter` or on a per-record basis as a column named topic in the dataframe.

In the following code example, we write key-value data from a dataframe into a specified Kafka topic:

```
query = streaming_kafka_df
  .selectExpr("CAST(user_id AS STRING) AS key", "to_
json(struct(*)) AS value") \
  .writeStream \
  .format("kafka") \
  .option("kafka.bootstrap.servers", "host1:port1,host2:port2")
\
  .option("topic", "topic1") \
  .option("checkpointLocation", path_to_HDFS).start()
```

The preceding query takes a dataframe containing user information and writes it to Kafka. `user_id` is a string used as the key. In this operation, we are transforming all the columns of the dataframe into JSON strings, putting the results in the value of the record.

In the next section, we will learn how to use Avro data in Structured Streaming.

Avro data

Apache Avro is a commonly used data serialization system. We can use Avro data in Azure Databricks using the `from_avro` and `to_avro` functions to build streaming pipelines that encode a column into and from binary Avro data. Similar to `from_json` and `to_json`, you can use these functions with any binary column, but you must specify the Avro schema manually. The `from_avro` and `to_avro` functions can be passed to SQL functions in streaming queries.

The following code snippet shows how we can use the `from_avro` and `to_avro` functions along with streaming data:

```
from pyspark.sql.avro.functions import from_avro, to_avro
streaming_df = spark
  .readStream
  .format("kafka")
  .option("kafka.bootstrap.servers", servers)
  .option("subscribe", "t")
  .load()
```

```
    .select(
        from_avro("key", SchemaBuilder.builder().stringType()).
as("key"),
        from_avro("value", SchemaBuilder.builder().intType()).
as("value"))
```

In this case, when reading the key and value of a Kafka topic, we must decode the binary Avro data into structured data, which has the following schema: <key: string, value: int>.

We can also convert structured data to binary from string and int, which will be interpreted as the key and value, and later on save it to a Kafka topic. The code to do this is illustrated in the following snippet:

```
streaming_df.select(
        to_avro("key").as("key"),
        to_avro("value").as("value"))
    .writeStream
    .format("kafka")
    .option("kafka.bootstrap.servers", servers)
    .option("article", "t")
    .save()
```

You can also specify a schema by using a JSON string to define the fields and its data types (for example, if "/tmp/userdata.avsc" is), as follows:

```
{
    "namespace": "example.avro",
    "type": "record",
    "name": "User",
    "fields": [
        {"name": "name", "type": "string"},
        {"name": " filter_value", "type": ["string", "null"]}
    ]
}
```

We can create a JSON string and use it in the `from_avro` function, as shown next. First, we specify the JSON schema to be read, as follows:

```
from pyspark.sql.avro.functions import from_avro, to_avro
json_format_schema = open("/tmp/ userdata.avsc", "r").read()
```

Then, we can use the schema in the `from_avro` function. In the following code example, we first decode the Avro data into a struct, filter by the `filter_value` column, and finally encode the name column in Avro format:

```
output = straming_df.
   .select(from_avro("value", json_format_schema).
alias("user"))\
   .where('user.filter_value == "value1"')\
   .select(to_avro("user.name").alias("value"))
```

The Apache Avro format is widely used in the Hadoop environment and it's a common option in data pipelines. Azure Databricks provides support for the ingestion and handling of sources using this file format.

The next section will dive into how we can integrate different data sinks in order to dump the data there, before and after it has been processed.

Data sinks

Sometimes, it is necessary to aggregate the receiving streams of data to be written into another location, which can be a data sink. Data sinks are a way in which we can call external sources where we store data and can be, for example, an S3 bucket where we want to keep copies of the aggregated streams of data. In this section, we will go through the steps in which we can write the streams of data into any external data sink location.

In Structured Streaming APIs, we have two ways to write the output of a query to external data sources that do not have an existing streaming sink. These options are the `foreachBatch` and `foreach` functions.

The foreachBatch method of writeStream allows existing batch data sources to be reused. This is done by specifying a function that is executed on the output data of every micro-batch of the streaming query. This function takes two parameters: the first parameter is the data frame that has the output data of a micro-batch, and the second one is the unique ID of that micro-batch. The foreachBatch method allows you to do the following:

- Reuse existing batch data sources.

- Write to multiple locations.

- Apply additional data frame operations.

In a case where there might not be a streaming sink available in the storage system but we can write data in batches, we can use foreachBatch to write the data in batches.

We can also write the output of a streaming query to multiple locations. We can do so by simply writing the output data frame into these multiple locations, although this might lead to data being computed every time we write into a different location and possibly having to read all the input data again. To avoid this, we can cache the output data frame, then write to the location, and finally uncache it.

In the following code example, we see how we can write to different locations using foreachBatch in Scala:

```
streaming_df.writeStream.foreachBatch { (batchDF: DataFrame,
batchId: Long) =>
  batchDF.persist()
  batchDF.write.format(format_location_1).save(your_location_1)
  batchDF.write.format(format_location_2).save(your_location_2)
  batchDF.unpersist()
}
```

The foreachBatch method helps us to overcome certain limitations regarding operations that are available for static DataFrames but not supported in streaming DataFrames. The foreachBatch option allows us to apply these operations to each micro-batch output.

The foreachBatch method guarantees data to be processed only once, although you can use the batchId parameter provided as a means to apply deduplication methods. It is important to also notice that one of the main differences between foreachBatch and foreach is that the former will not work for continuous processing modes as it fundamentally relies on micro-batch execution, while the latter gives an option to write data in continuous mode.

If, for some reason, `foreachBatch` is not an available option to use, then you can code your custom writer using the `foreach` option and have the data-writing logic be divided into `open`, `process`, and `close` methods.

In Python, we can make use of the `foreach` method either in a function or in an object. The function provides an effective way to code your processing logic but will not allow deduplicating of output data in the case of failure, causing the reprocessing of some input data. In that case, you should specify the processing logic within an object.

The following custom function takes a row as input and writes the row to storage:

```python
def process_row (row):
    """
    Custom function to write row to storage.
    """
    pass
```

Then, we can use this function in the `foreach` method of `writeStream`, as follows:

```python
query = streaming_df.writeStream.foreach(process_row).start()
```

To construct the object, it needs to have a process method that will write the row to the storage. We can also add optional `open` and `close` methods to handle the connection, as illustrated in the following code snippet:

```python
class ForeachWriter:
    def open(self, partition_id, epoch_id):
    """
    Optional method to open the connection
    """
    def process(self, row):
        """
        Custom function to write row to storage. Required method.
        """
    def close(self, error):
    """
    Optional method to close the connection
    """
```

Finally, we can call this object on the query using the `foreach` method of `writeStream`, as illustrated in the following code snippet:

```
query = streaming_df.writeStream.foreach(ForeachWriter()).
start()
```

In this way, we can very easily define a class with methods that process the streams of data we receive using Structured Streaming and then write to external locations that will be used as data sinks of the processed streams of data, using the `writeStream` method.

In the next section, we will go through the steps that we should follow in order to maintain integrity of the data, regardless of query failures or interruptions of the stream, by specifying the use of checkpoints.

Recovering from query failures

Failures can happen because of changes in the input data schema, changes along tables used in the computation, missing files, or due to many other root causes. When developing applications that make use of streams of data, it is necessary to have robust failure-handling methods. Structured Streaming guarantees that the results will remain valid even in the case of failure. To do this, it places two requirements on the input sources and output sinks, outlined as follows:

- The input sources should be replayable, which means that recent data can be read again if the job fails.
- The output sinks should allow transactional updates, meaning that these updates are atomic (while the update is running, the system should keep functioning) and can be rolled back.

One of the previously mentioned features of Structured Streaming is checkpointing, which can be enabled for streaming queries. This also allows a stream to quickly recover from a failure by simply restarting the stream after a failure, which will cause the query to continue from where it was left off, while guaranteeing data consistency. So, it is always recommended to configure **Enable Query Checkpointing** and to configure **Databricks Jobs** to restart your queries automatically after a failure.

To enable checkpointing, we can set the `checkpointLocation` option to the desired checkpoint path when writing the dataframe. An example of this is shown in the following code snippet:

```
streaming_df.writeStream
    .format("parquet")
```

```
   .option("path", output_path)
   .option("checkpointLocation", checkpoint_path)
   .start()
```

This checkpoint location preserves all information about a query. Therefore, all queries must have a unique checkpoint path.

We face limitations on the possible changes in a streaming query that are allowed between restarts using the same checkpoint path.

By saying that an operation is allowed, it is understood that we can implement the change, but whether the semantics or its effects will be as desired will depend on the query and the specific change applied.

The term *not allowed* means the change will likely fail when the query is restarted.

Here is a summary of the types of changes we can make:

- We are not allowed to change the number or type of input sources.
- If we can change parameters of input sources, this will depend on the source and query.
- Some changes in the type of output sink are allowed.
- If we can make changes to the parameters of the output sink, this will depend on the type of sink and query.
- Some changes are allowed on operations, such as map and filter operations.
- Changes in stateful operations can lead to failure due to changes in the schema of state data needed to update the result.

As we can see, Structured Streaming provides several features in order to provide fault tolerance for production-grade processing of streams of data.

In the next section, we will gain knowledge on how we can optimize the performance of our streaming queries.

Optimizing streaming queries

We can have a streaming query that makes use of stateful operations such as streaming aggregation, streaming dropDuplicates, mapGroupsWithState, or flatMapGroupsWithState, streaming to stream joins, and so on. Running these operations continuously results in a need to maintain millions of keys of the state of the data in the memory of each executor, leading to bottlenecks.

To overcome this issue, Azure Databricks provides a solution, which is to keep the state of the data in a RocksDB database instead of using the executor memory. RocksDB is a library that provides a persistent key-value store to efficiently manage the state of data in the memory and the local SSD. Therefore, used in combination with Structured Streaming checkpoints, these solutions guarantee safeguards against failures.

You can enable RocksDB-based state management by setting the following configuration in SparkSession before starting the streaming query:

```
spark.conf.set(
  "spark.sql.streaming.stateStore.providerClass",
  "com.databricks.sql.streaming.state.
RocksDBStateStoreProvider")
```

RocksDB can maintain 100 times more state keys than the standard executor memory. We can also improve the performance of our queries using compute-optimized instances such as Azure Standard_F16s instances as workers and setting the number of shuffle partitions to up to two times the number of cores in the cluster.

Triggering streaming query executions

Triggers are a way in which we define events that will lead to an operation being executed on a portion of data, so they handle the timing of streaming data processing. These triggers are defined by intervals of time in which the system checks if new data has arrived. If this interval of time is too small this will lead to unnecessary use of resources, so it should always be an amount of time customized according to your specific process.

The parameters of the triggers of the streaming queries will define if this query is to be executed as a micro-batch query on a fixed batch interval or as a continuous processing query.

Different kinds of triggers

There are different kinds of triggers available in Azure Databricks that we can use to define when our streaming queries will be executed. The available options are outlined here:

- **Unspecified trigger**: This is the default option and means that unless specified otherwise, the query will be executed in micro-batch mode, where micro-batches are generated when the processing previous micro-batch has already been executed successfully.

- **Triggers based on fixed-interval micro-batches**: The query will be executed with micro-batches mode, where micro-batch execution will be started at specified intervals. If the previous micro-batch execution is completed in less than the specified interval, Azure Databricks will wait until the interval is over before kicking off the next micro-batch. If the execution of the micro-batch takes longer than the specified interval to complete, then the next batch will be executed as soon as the previous one is done.

  If no new data is available, no micro-batch will be executed.

- **One-time micro-batch triggers**: The query will run just once to process all the available data. This is a good option when clusters are started and turned off to save resources, and therefore we can just process all the new data available since the start of the cluster.

- **Continuous trigger with fix checkpoint interval**: The query will be kicked off in a continuous low-latency processing mode.

Next, we will see some examples of how we can apply these different types of triggers when using Structured Streaming for streaming data.

Trigger examples

The following code snippet provides an example of the default trigger in Structured Streaming, which—as mentioned before—runs the micro-batch as soon as it can:

```
streaming_df.writeStream \
  .format("console") \
  .start()
```

We can also define a specific interval of time, making use of the `processingTime` trigger option. In the following code example, we set a 10-second micro-batch interval of time:

```
streaming_df.writeStream \
  .format("console") \
  .trigger(processingTime='10 seconds').start()
```

As mentioned before, we can set a one-time trigger to process all of our new data just once, as illustrated in the following code snippet:

```
streaming_df.writeStream \
  .format("console") \
  .trigger(once=True).start()
```

Finally, we can set a continuous trigger that will have a 2-second checkpointing interval of time for saving data, to recover from eventual failures. The code for this is illustrated in the following snippet:

```
streaming_df.writeStream
  .format("console")
  .trigger(continuous='1 second').start()
```

Triggers in Structured Streaming allow us to have fine-grained control over how our query will behave when processing new data that arrives in defined intervals of time.

In the last section of this chapter, we will dive into which tools are available in Azure Databricks to visualize data.

Visualizing data on streaming data frames

When working with streams of data in Structured Streaming data frames, we can visualize real-time data using the `display` function. This function is different from other visualizing functions because it allows us to specify options such as `processingTime` and `checkpointLocation` due to the real-time nature of the data. These options are set in order to manage the exact point in time we are visualizing and should be always be set in production in order to know exactly the state of the data that we are seeing.

In the following code example, we first define a Structured Streaming dataframe, and then we use the `display` function to show the state of the data every 5 seconds of processing time, on a specific checkpoint location:

```
streaming_df = spark.readStream.format("rate").load()
display(streaming_df.groupBy().count(), processingTime = "5
seconds", checkpointLocation = "<checkpoint-path>")
```

Specifically, these are the available parameters for this function:

- The `streamName` parameter is the specific streaming query in question that we want to visualize.

- The `processingTime` parameter defines the interval of time for how often the streaming query is executed. If it's not specified, Azure Databricks will check for the availability of new data when the previous process has been completed—a behavior that could lead to undesired cost increases. Therefore, this is an option that is recommended to be set explicitly when possible.

The `checkpointLocation` parameter is the path to where the checkpoint data is written. If not specified, the location will be temporary. It is recommended to set this parameter so that a stream can be restarted from where it was left in the case of any failure.

The Azure Databricks `display` function supports a variety of plot types. We can choose different types of plots by selecting the desired type of plot from the drop-down menu in the top-left corner of the cell where we executed the function, as illustrated in the following screenshot:

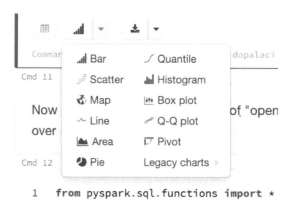

Figure 6.1 – Available options for the display function

We have an option to create different types of plots such as bar, scatter, and line plots, with the extra feature of being able to set options for these visualizations. If we click on the **Plot Options** menu, we will be prompted with a window where we can customize the plot depending on the selected type, as illustrated in the following screenshot:

Figure 6.2 – Available options for customizing a plot

The `display` option provides ease of use when visualizing data in general and streaming queries in particular. The provided options allow us to easily drag and drop the variables required to create each type of plot, selecting the type of aggregation executed and modifying several aspects of its design.

The next section will comprise of an example in which we will simulate a stream of data and use Structured Streaming to run streaming queries.

Example on Structured Streaming

In this example, we will be looking at how we can leverage knowledge we have acquired on Structured Streaming throughout the previous sections. We will simulate an incoming stream of data by using one of the example datasets in which we have small JSON files that, in real scenarios, could be the incoming stream of data that we want to process. We will use these files in order to compute metrics such as counts and windowed counts on a stream of timestamped actions. Let's take a look at the contents of the `structured-streaming` example dataset, as follows:

```
%fs ls /databricks-datasets/structured-streaming/events/
```

You will find that there are about 50 JSON files in the directory. You can see some of these in the following screenshot:

Figure 6.3 – The structured-streaming dataset's JSON files

We can see what one of these JSON files contains by using the `fs head` option, as follows:

```
%fs head /databricks-datasets/structured-streaming/events/file-
0.json
```

We can see that each line in the file has two fields, which are `time` and `action`, so our next step will be to try to analyze these files interactively using Structured Streaming, as follows:

1. The first step that we will take is to define the schema of our data and store it in a JSON file that we can later use to create a streaming dataframe. The code for this can be seen in the following snippet:

    ```
    from pyspark.sql.types import *
    input_path = "/databricks-datasets/structured-streaming/
    events/"
    ```

2. Since we already know the structure of the data we won't need to infer the schema, so we will use PySpark SQL types to define the schema of our future streaming dataframe, as illustrated in the following code snippet:

    ```
    json_schema = StructType([StructField("time",
                                          TimestampType(),
                                          True),
                              StructField("action",
                                          StringType(),
                                          True)])
    ```

3. Now, we can build a static data frame using the schema defined before and loading the data in the JSON files, as follows:

    ```
    static_df = (
      spark
        .read
        .schema(json_schema)
        .json(input_path)
    )
    ```

4. Finally, we can visualize the data frame using the `display` function, as mentioned in the previous section, as follows:

    ```
    display(static_df)
    ```

The output is as follows:

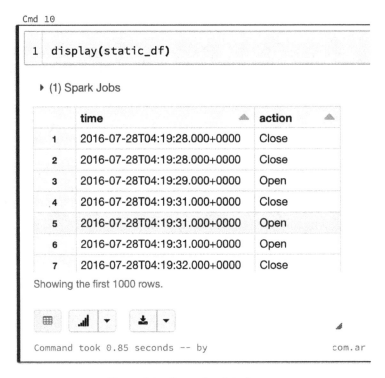

Figure 6.4 – Streaming data frame

5. We are now ready to run the computation in order to know the amount of open and close orders by grouping data based on windowed aggregations of 1 hour. We will import everything from PySpark SQL functions using an asterisk (*) to get, among others, the window function, as illustrated in the following code snippet:

```
from pyspark.sql.functions import *
static_count_df = (
  static_df
    .groupBy(
      static_df.action,
      window(static_df.time, "1 hour"))
    .count()
)
static_count_df.cache()
```

6. Next, we register the dataframe as a temporary view called `data_counts`, as illustrated in the following code snippet:

```
static_count_df.createOrReplaceTempView("data_counts")
```

7. Finally, we can directly query the table using SQL commands. For example, to compute the total counts throughout all the hours, we can run the following SQL command:

```
%sql
select action, sum(count) as total_count from data_counts
group by action
```

The output is as follows:

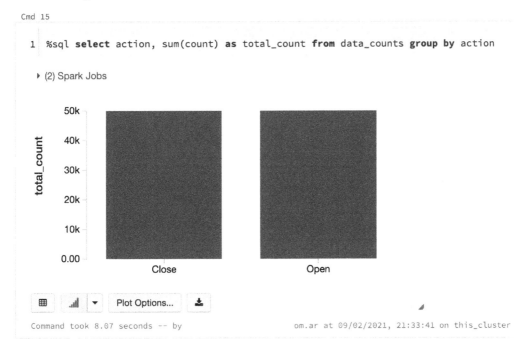

Figure 6.5 – SQL queries on a streaming data frame

8. We can also use windowed counts to create a timeline grouped by `time` and `action`, as follows:

```
%sql
select action, date_format(window.end, "MMM-dd HH:mm") as
time, count from data_counts order by time, action
```

The output is as follows:

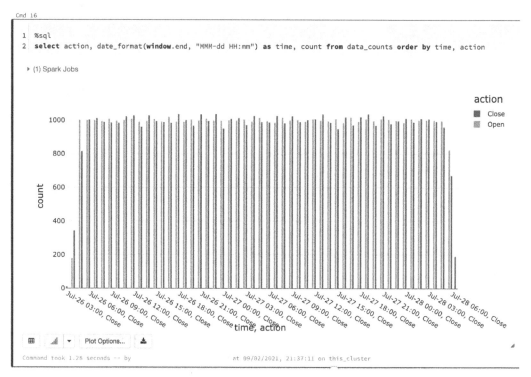

```
Cmd 16
1  %sql
2  select action, date_format(window.end, "MMM-dd HH:mm") as time, count from data_counts order by time, action
```

▶ (1) Spark Jobs

Command took 1.28 seconds -- by   at 09/02/2021, 21:37:11 on this_cluster

Figure 6.6 – Windowed counts on a streaming data frame

9. In the following code example, we will use a similar definition to the one used here but using the readStream option instead of the read method. Also, in order to simulate a stream of files, we will pick one file at a time using the maxFilesPerTrigger option and setting it to 1:

```python
from pyspark.sql.functions import *
streaming_df = (
  spark
    .readStream
    .schema(json_schema)
    .option("maxFilesPerTrigger", 1)
    .json(input_path)
)
```

10. We can then make the same query that we have applied to the static dataframe previously, as follows:

```
streaming_counts_df = (
  streaming_df
    .groupBy(
      streaming_df.action,
      window(streaming_df.time, "1 hour"))
    .count()
)
```

11. We can check that the data frame is actually a streaming source of data by running the following code:

```
streaming_df.isStreaming
```

The output is as follows:

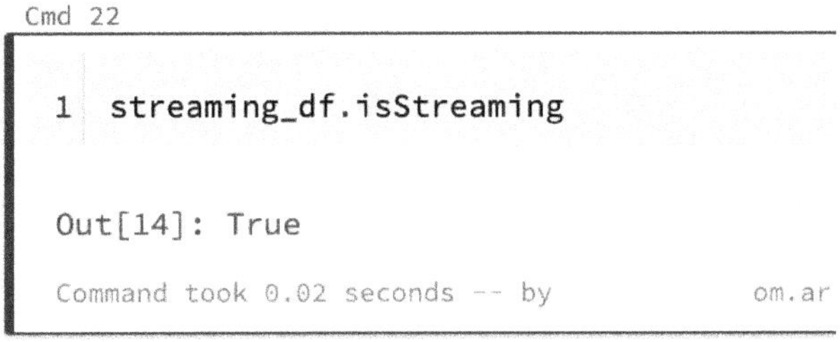

```
Cmd 22

1  streaming_df.isStreaming


Out[14]: True

Command took 0.02 seconds -- by          om.ar
```

Figure 6.7 – Checking if the dataframe is a streaming source

As you can see, the output of streamingCountsDF.isStreaming is True, therefore streaming_df is a streaming dataframe.

12. The next option is set, to keep the size of shuffles small, as illustrated in the following code snippet:

```
spark.conf.set("spark.sql.shuffle.partitions", "2")
```

13. We can then store the results of our query, passing as format "memory" to store the result in an in-memory table. The queryName option will set the name of this table to data_counts. Finally, we set the output mode to complete so that all counts are on the table. The code can be seen in the following snippet:

```
query =
  streaming_counts_df
    .writeStream
```

```
    .format("memory")
    .queryName("data_counts")
    .outputMode("complete")
    .start()
```

The `query` object is a handle that runs in the background, continuously looking for new files and updating the windowed resulting aggregation counts.

Note the status of query in the preceding code snippet. The progress bar in the cell shows the status of the query, which in our case is active.

14. We will artificially wait some time for a few files to be processed and then run a query to the in-memory `data_counts` table. First, we introduce a sleep of 5 seconds by running the following code:

```
from time import sleep
sleep(5)
```

15. Then, we can run a query in another cell as a SQL command, using SQL magic, as follows:

```
%sql
select action, date_format(window.end, "MMM-dd HH:mm") as
time, count
from data_counts
order by time, action
```

The following screenshot shows us the resulting graph created by running this query in one of the Databricks notebook code cells:

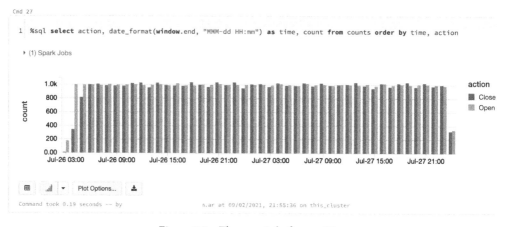

Figure 6.8 – The counts before waiting

We can see the timeline of windowed counts growing bigger. Running these queries will always result in the latest updated counts, which the streaming query is updating in the background.

16. If we wait a few seconds more, we will have new data computed into the result table, as illustrated in the following code snippet:

```
sleep(5)
```

17. And then, we run the query again, as follows:

```
%sql
select action, date_format(window.end, "MMM-dd HH:mm") as time, count
from data_counts
order by time, action
```

We can see in the following screenshot of a graph created by running the query that the results at the right side of the graph have changed as more data is being ingested:

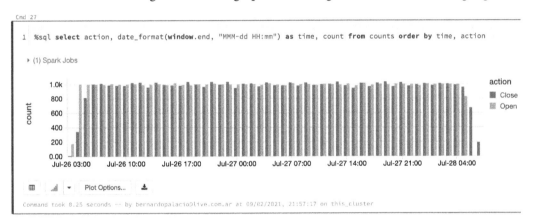

Figure 6.9 – The counts after waiting

18. We can also see the resulting number of `"opens"` and `"closes"` by running the following SQL query:

```
%sql
select action, sum(count) as total_count
from data_counts
group by action order by action
```

This example has its limitations due to the small number of files present in the dataset. After consuming them all, there will be no more updates to the table.

19. Finally, we can stop the query running in the background, either by clicking on the **Cancel** link in the cell of the query or by executing the `query.stop` function, as illustrated in the following code snippet:

```
query.stop()
```

Either way, the status of the cell where the query is running will be updated to TERMINATED.

Summary

Throughout this chapter, we have reviewed different features of Structured Streaming and looked at how we can leverage them in Azure Databricks when dealing with streams of data from different sources.

These sources can be data from Azure Event Hubs or data derived using Delta tables as streaming sources, using Auto Loader to manage file detection, reading from Apache Kafka, using Avro format files, and through dealing with data sinks. We have also described how Structured Streaming provides fault tolerance while working with streams of data and looked at how we can visualize these streams using the display function. Finally, we have concluded with an example in which we have simulated JSON files arriving in the storage.

In the next chapter, we will dive more deeply into how we can use the PySpark API to manipulate data, how we can use Python popular libraries in Azure Databricks and the nuances of installing them on a distributed system, how we can easily migrate from Pandas into big data with the Koalas API, and—finally—how to visualize data using popular Python libraries.

Section 3: Machine and Deep Learning with Databricks

This section explores concepts around the use of machine learning and deep learning models. At the end of this section, you will be able to create, train, and deploy predictive models harnessing the power of distributed computing.

This section contains the following chapters:

- *Chapter 7, Using Python Libraries in Azure Databricks*
- *Chapter 8, Databricks Runtime for Machine Learning*
- *Chapter 9, Databricks Runtime for Deep Learning*
- *Chapter 10, Model Tracking and Tuning in Azure Databricks*
- *Chapter 11, Managing and Serving Models with MLflow and MLeap*
- *Chapter 12, Distributed Deep Learning in Azure Databricks*

7
Using Python Libraries in Azure Databricks

Azure Databricks has implementations on different programming languages, but we will focus on Python developers, therefore we will explore all the nuances regarding working with it, as well as introducing core concepts regarding models and data that later will be studied in more detail.

In this chapter, we will cover the following:

- Installing popular Python libraries in Azure Databricks
- Learning key concepts of the PySpark API
- Using the Koalas API to manipulate data in a similar way as we would do with pandas
- Using visualization libraries to make plots and graphics

These concepts will be introduced more deeply in the next sections of this chapter.

Technical requirements

This chapter will require you to have an Azure Databrick subscription available to work on the examples in notebooks attached to running clusters.

Without further ado, we can start by looking into the different ways in which we can install libraries in Azure Databricks and the differences of each method.

Installing libraries in Azure Databricks

We can make use of third-party or custom code by installing libraries written in Python, Java, Scala, or R. These libraries will be available to notebooks and jobs running on your clusters depending on the level at which the libraries were installed.

In Azure Databricks, installing libraries can be done in different ways, the most important decision being at which level we will be installing these libraries. The options available are at the workspace, cluster, or notebook level:

- **Workspace** libraries serve as a local repository from which you create cluster-installed libraries. A workspace library might be custom code created by your organization or might be a particular version of an open-source library that your organization has standardized on.

- **Cluster** libraries are available to be used by all notebooks attached to that cluster. You can install a cluster library from a public repository or create one from a previously installed workspace library.

- **Notebook-scoped**: Python libraries allow you to install Python libraries and create an environment scoped to a notebook session. Notebook-scoped libraries will not be visible in other notebooks running on the same cluster. These libraries do not persist and need to be installed at the beginning of each session. Use notebook-scoped libraries when you need a custom Python environment for a specific notebook. With these libraries, it is easy to create, modify, save, reuse, and share Python environments. Notebook-scoped libraries are available using `%pip` in all Databricks Runtime instances and using the `%conda` magic commands in Databricks Runtime ML 6.4.

Azure Databricks includes many popular Python libraries already installed. In order to check whether the library we want to install is already available, we can see which libraries are included in Databricks Runtime by looking at the **System Environment** subsection of the **Databricks Runtime** release notes for your Databricks Runtime version.

Workspace libraries

Workspace libraries act as a local repository from where cluster-installed libraries can be created. These libraries can be custom code created by you or your company or organization; it can be a particular version of an open-source library that has been standardized or simply might hold features that are necessary for the job you are doing.

We must first install the workspace library on a cluster before it can be made available to notebooks or jobs running on that cluster. We can create a workspace library in the `Shared` folder, which will be made available to all users in that workspace, or we can create it in a user folder, which will make the library available only to that user.

To install a library, right-click on the workspace folder where you want to store the library and select the **Create | Library** option:

Figure 7.1 – Creating a library in a directory

You will be prompted with the **Create Library** dialog from where you need to select the required library source and follow the appropriate procedure.

Figure 7.2 – Creating a library from the UI

The available options are as follows:

- **Upload** a library.
- Reference an uploaded library.
- **PyPI** package.
- **Maven** package.
- **CRAN** package (for R packages):

To install a workspace library, we can use a Jar, Python egg, or Python wheel. To do this, drag your library's Egg or wheel to the drop box, optionally enter a library name, and then click on the **Create** button. The file will be uploaded to **Databricks File System** (**DBFS**) and we can also optionally install the library on a cluster. You can also reference Eggs or wheels already in the workspace or in an S3 bucket.

Installed libraries in the running clusters, along with their details and status, are shown in the workspace by clicking on the library details.

If you want to remove a workspace library, you must first uninstall it before you can delete the directory that contains the library in the workspace by simply moving the library to the `Trash` folder after it was uninstalled from the cluster.

Cluster libraries

Cluster libraries can be used by all notebooks and jobs running on that cluster. These libraries can be installed directly from a public repository such as `PyPI` or `Maven`, from a previously installed workspace library, or using an `init` script.

There are three ways to install a library on a cluster:

- Install a workspace library that has been already been uploaded to the workspace.
- Install a library for use with a specific cluster only.
- Using an `init` script that runs at cluster creation time.

Bear in mind that if a library requires a custom configuration, you will need to install it using an `init` script that runs at cluster creation time. Also consider that when a library is installed on a cluster, notebooks and jobs attached to that cluster will not immediately see the new library. Therefore, you must first detach and then reattach the notebook to the cluster in order to make it visible.

Azure Databricks will process all workspace libraries in the order in which they were installed on the cluster, so it's also good to pay attention to these situations if there are dependencies between libraries and the order of installation of the required libraries is relevant.

In order to install a library that already exists in the workspace, you can start either from the cluster UI or the library UI:

- **Cluster UI**: To install using the cluster UI, click the **Clusters** option on the sidebar, select the cluster where you want to install the library, and in the **Libraries** tab, click on **Install New**. In the **Library Source** button list, select **Workspace**, then select a workspace library, and finally, click **Install**. You can also configure the library to be installed on all clusters by clicking on the library and using the Install **automatically on all clusters** checkbox option.

- **Library UI**: To install a library using the UI, go to the folder containing the library and then click on the library name and follow the steps mentioned in the previous section, selecting the option to install on the cluster:

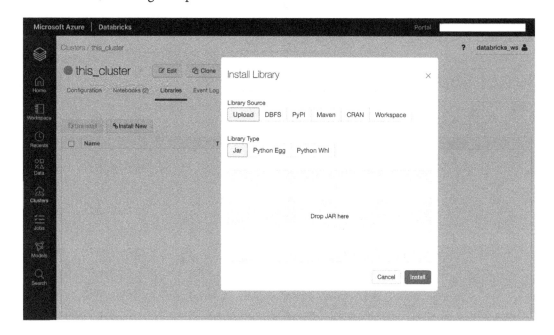

Figure 7.3 – Installing a cluster-scoped library from the UI

As mentioned before, if your library requires a custom configuration, you will need to install the library using an `init` script.

To uninstall a library, you can start from a cluster or a library, go to the cluster UI, select the cluster where the library is, and in the **Libraries** tab, select the checkbox next to the cluster you want to uninstall the library from, then click **Uninstall** then **Confirm**. The status will change to **Uninstall pending restart**.

To uninstall using the library UI, go to the folder containing the library, click the library name, and select the checkbox next to the cluster you want to uninstall the library from, then click **Uninstall** then **Confirm**. The status changes to **Uninstall pending restart**.

In the same way as during the installation, to uninstall a library from a cluster, the library is removed only when your cluster is restarted. Until then, the status of the uninstalled library will be shown as **Uninstall pending restart**.

To update a library installed on the cluster, you should uninstall the version you want to update and install the new one.

Notebook-scoped Python libraries

We can specify libraries to be installed at the notebook level if we don't want to affect all the notebooks attached to a cluster by installing, for example, a custom or untested library to it. Notebook-scoped allows you to manage and share custom Python environments running in a specific notebook. When we install a library at the notebook level, we only affect the notebook and any jobs associated with it, which will be the only ones able to access that library, and therefore, other notebooks attached to the same cluster will not be affected.

All notebook-scoped libraries do not persist across sessions and must be reinstalled at the beginning of each session or when the notebook is disconnected from the computing cluster.

There are three methods for installing notebook-scoped libraries:

- Run the `%pip` magic command in a notebook. The `%pip` command is supported on all versions of Databricks Runtime from 7.1 and above.

- The `%conda` command, which is only supported, on Databricks Runtime 6.4 ML and above and Databricks Runtime 6.4 for Genomics and above.

- We can use Databricks **Library** utilities, which allow us to install Python libraries and create an environment scoped to a notebook session, available both on the driver and on the executors. The **Library** utilities are supported only on Databricks Runtime, not Databricks Runtime ML or Databricks Runtime for Genomics.

One important thing to stress is that the `%conda` magic commands are not available on Databricks Runtime. They are only available on Databricks Runtime ML and Databricks Runtime for Genomics. In these environments, we can use `%conda` in the same way as other Python environments, except for the following `conda` commands, which are not supported:

- `activate`
- `create`
- `init`
- `run`
- `env create`
- `env remove`

Using %conda allows us to save, reuse, or share an environment. The environment is lost every time that a notebook is detached from a cluster, but we can use %conda to save it to be used or shared later on. Databricks recommends that these environments are executed in clusters running the same version of Databricks. We can save the environment as a conda YAML specification:

```
%conda env export -f /dbfs/sub_dir/my_env.yml
```

Import the file to another notebook using conda env update:

```
%conda env update -f dbfs/sub_dir/my_env.yml
```

While in Databricks Runtime 6.4 ML and above and Databricks Runtime 6.4 for Genomics, we can use %pip and %conda indistinctly; using them in combination can cause interactions between commands. To avoid conflicts, it is important to keep the next statements in mind when working with %pip and %conda:

- If you have installed libraries using the API or the cluster UI, you should use only %pip commands when installing libraries at the notebook level.

- If you use notebook-scoped libraries on a cluster, init scripts running on the cluster to which the notebook is attached can use either conda or pip commands to install libraries. However, if the init script uses pip commands, you should only use %pip commands in notebooks and not %conda to keep consistency across versions.

- It's always best to pick either pip commands or conda commands to manage libraries and not a combination of both.

Because our focus is on Databricks Runtime, we will focus on using %pip as the default method to manage libraries at the notebook level.

When we install libraries at the notebook level, the driver node in our cluster still needs to install these libraries across nodes. Therefore, using notebook-scoped libraries will create more traffic between driver and nodes, as the driver tries to keep the environment consistent across executor nodes. When working with clusters that have more than 10 nodes, the minimum specifications recommended by Databricks for the driver node to keep up with the orchestration are as follows:

- Use Lsv2 storage-optimized instance types when using a 100-node CPU cluster.

- Use accelerated compute and high-performance instance types such as NC or NV for a 10-node GPU cluster.

- For larger clusters, use a driver node with larger processing specifications.

The notebook state is reset after any %pip or %conda command that modifies the environment, so it is important to consider that all these commands must be at the beginning of the notebook in order to not lose created variables and lose unsaved results. So, if we create methods, functions, or variables and then we execute either the %pip or %conda commands, these methods and variables won't be recovered.

Sometimes, as any experienced Python user will know, the use of %pip or %conda to install or uninstall packages can lead to some unstable behavior, or even cause some features to stop working. In order to overcome this issue, one of the simplest solutions is to restart the environment by detaching the notebook from the running cluster and then re-attaching it back. If this doesn't solve the issue, we can directly restart the cluster to which the notebook is attached in order to make the driver and executor nodes reinstall all the libraries from a clean state:

- Using %pip or %conda to install or uninstall packages is equivalent to its use in any other Python environment. For example, to install a specific version of a library, we can run the following on a cell:

```
%pip install pandas==1.2.1
```

- Or, on the other hand, we can install a Wheel package by using %pip install and the path to where our package is located:

```
%pip install /path/to/my_package.whl
```

- To uninstall a library, it is necessary to pass the -y option to say yes as the default, because there is no keyboard input when running this option:

```
%pip uninstall -y pandas
```

Although we can install and uninstall libraries using the %pip command, we cannot uninstall libraries included in the Databricks Runtime or any libraries that are installed at the cluster level. If you have installed in your notebook a library that has a version that is different from the one installed in Databricks Runtime or in the cluster, using %pip to uninstall it will revert the version to the one that is installed in Databricks Runtime or in the cluster:

- We can also install a library from a repository such as GitHub with the %pip command, specifying options for the version or Git subdirectory:

```
%pip install git+https://github.com/databricks/
databricks-cli
```

- You can also install a custom package from DBFS with the `%pip` command. To do this, we need to run the `%pip install` command and specify the path in DBFS where our custom package is located:

```
%pip install /dbfs/mypackage-0.0.1-py3-none-any.whl
```

- This option allows us to install packages locally using Python Wheels. It is important to notice that when files are uploaded to DBFS, this will automatically rename the file, replacing spaces, periods, and hyphens with underscores. This can be an issue because `pip` expects that the Wheel file will use periods to express the version (such as version 1.0.1) and hyphens instead of spaces or underscores. Therefore, to install a Python package from a Wheel uploaded to DBFS, we must first rename the file to make it consistent to what `pip` is expecting:

```
%pip install /dbfs/mycustompackage-0.0.1-py3-none-any.whl
```

- We can also use `%pip` to list all the installed libraries in a requirements file:

```
%pip freeze > /dbfs/sub_dir/requirements.txt
```

Any subdirectories included in the file path should be already created before running `%pip freeze`. If you run `%pip freeze` and the directory is not created properly with all its subfolders, the command will fail.

- We can also use `%pip` to install libraries using a requirements file. A requirements file contains a list of Python packages to be installed using `pip`, such as the one we have just created using `%pip freeze`, and it is always a text file named `requirements`. An example of how we can install libraries using a requirements file located in DBFS is as follows:

```
%pip install -r /dbfs/requirements.txt
```

This way, we can install all the required libraries without having to install each of them manually and this helps us reproduce environments to recreate results and processes.

Some final remarks about managing libraries at the notebook level are as follows:

- Libraries installed using an `init` script will be accessible by all notebooks attached to that cluster.
- Libraries installed from the cluster UI or API are available to all notebooks on the cluster, which are installed using `pip` and therefore you should use only `%pip` commands in those notebooks.

- Clusters running Databricks Runtime ML or Databricks Runtime for Genomics can use `init` scripts running either `conda` or `pip` commands to install libraries. However, if the `init` script only includes `pip` commands, then you should use only `%pip` commands in the notebooks.

Managing libraries at the notebook level allow us to provide a separation of concerns when running experiments without affecting all the notebooks attached to a cluster. It is important to understand how this relates to libraries located at the cluster or workspace level and how we can leverage this to efficiently manage libraries.

In the next section, we will discuss more deeply how to use the PySpark API in Azure Databricks.

PySpark API

We have been using the PySpark API across all sections when describing the features of Azure Databricks without discussing too much of its functionalities and how we can leverage them to make reliable ETL operations when working with big data. PySpark is the Python API for Apache Spark, a cluster-computing framework that is the heart of Azure Databricks.

Main functionalities of PySpark

PySpark allows you to harness the power of distributed computing with the ease of use of Python and it's the default way in which we express our computations through this book unless stated otherwise.

The fundamentals of PySpark lies in the functionality of its sub-packages of which the most central are the following:

- **PySpark DataFrames**: Data stored in rows following a set of named columns. These DataFrames are immutable and allow us to perform lazy computations.

- **The PySpark SQL module**: A higher-abstraction module for processing structured and semi-structured datasets from various sources and formats.

- **Streaming source**: A fault-tolerant and scalable module to read streaming data, which is converted into batches and read in predefined intervals of time, processes it using complex algorithms expressed in functions such as map, reduce, and so on, and then finally writes to different filesystems.

- **MLlib**: A library with several cluster-computing optimized machine learning algorithms for classification, regression, clustering, data preparation, utilities, and more.

- **RDDs**: This stands for **Resilient Distributed Datasets**. This is a layer of abstraction that sits on top of a distributed set of data and is the building block for any processes running in Azure Databricks.

Because our focus will be placed on working with data, it is very important for us to understand PySpark DataFrames; therefore, we will review the basics of operating with them in the next section.

Operating with PySpark DataFrames

In this section, we will introduce common PySpark DataFrame functions using Python that can be executed in any Azure Databricks notebook. The steps are as follows:

1. Our first step is to create a PySpark DataFrame. We can do so by using the `createDataFrame` method to create a DataFrame named `df` with columns named `id` and `value` in the following way:

    ```
    data = [(1, 'c1', 'a1'), (2, 'c2', 'a2')]
    columns = ['id', 'val1', 'val2']
    df = spark.createDataFrame(
        data,
        columns)
    ```

    When using the `createDataFrame` method, we must take into account the following:

    a) If the schema is defined by a list of column names, the type of each column will be inferred from data.

    b) If the schema is defined using `pyspark.sql.types.DataType` or a datatype string, it must match the real data.

2. For example, to create a blank DataFrame, we can first define the schema of data using `StructField` and store the information in a variable named `schema`:

    ```
    columns = [
        StructField("id", IntegerType(), True),
        StructField("value", StringType(), True),
    ]
    schema = StructType(columns)
    ```

3. After this, we can create the empty DataFrame using `createDataFrame` and `sc.emptyRDD`, along with the defined data schema:

```
multiplier_df = sqlContext.createDataFrame(sc.emptyRDD(),
schema)
```

4. Then, to add a row to it, we can insert them as the union of a temporary DataFrame:

```
row = [(1, "c1","a1")]
rdd = sc.parallelize(row)
df_temp = spark.createDataFrame(rdd, schema)
```

5. Finally, we can use the `union` function to join row-wise both DataFrames:

```
df = df.union(df_temp)
```

6. We can also create DataFrames from lists of data using the `createDataFrame` and `toDF` methods:

```
df = spark.createDataFrame(data).toDF(*columns)
```

7. Of course, we can also create a PySpark DataFrame from CSV and JSON files. For example, to read the data from a CSV into a DataFrame, we can do the following:

```
df = spark.read.csv("/dbfs/file_path.csv")
```

8. Similarly, we can do the same with a JSON file:

```
df = spark.read.csv("/dbfs/file_path.json")
```

9. And the same applies to TXT files:

```
df = spark.read.csv("/dbfs/file_path.txt")
```

10. After we have loaded our data into a DataFrame, we are ready to start operating with it. One of the most common operations to do in our data is to perform a union with other DataFrames with the same schema, as we have seen before when we used this method to append a new row to an existing DataFrame. For example, we can create a DataFrame:

```
data = [(1, 'c1', 'a1'), (2, 'c2', 'a2')]
columns = ['id', 'val1', 'val2']
df =spark.createDataFrame(
    data,
```

```
        columns
)
```

Then, we can display the schema and show the DataFrame:

```
df.printSchema()
df.show(truncate=False)
```

This results in the following output:

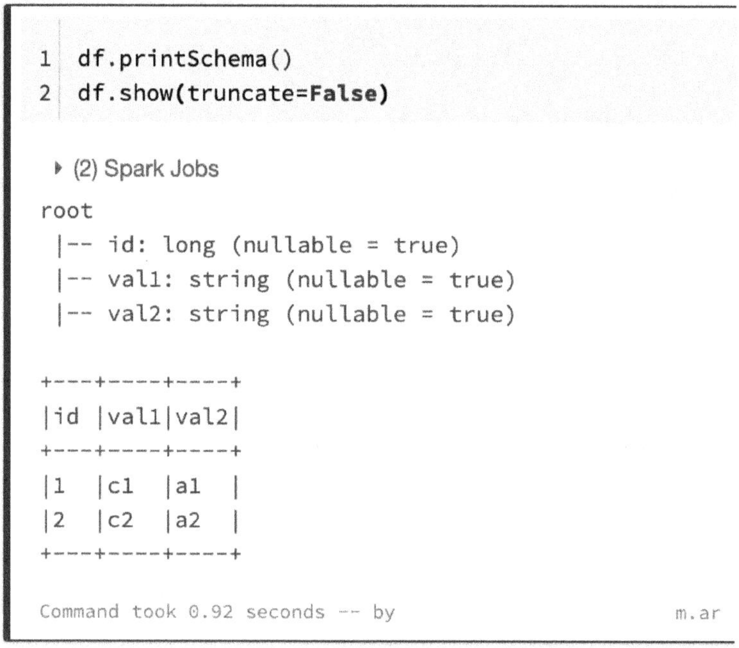

```
1  df.printSchema()
2  df.show(truncate=False)


▶ (2) Spark Jobs
root
 |-- id: long (nullable = true)
 |-- val1: string (nullable = true)
 |-- val2: string (nullable = true)


+---+----+----+
|id |val1|val2|
+---+----+----+
|1  |c1  |a1  |
|2  |c2  |a2  |
+---+----+----+

Command took 0.92 seconds -- by                         m.ar
```

Figure 7.4 – The schema of the DataFrame, along with its content

Then, we create another DataFrame, which we will merge with the previously created one, considering that they must have the same schema:

```
data_new = [(3, 'c2', 'a2'), (4, 'c3', 'a3')]
columns_new = ['id', 'val1', 'val2']
df_new =spark.createDataFrame(
    data_news,
    columns_new
)
```

Then, we display the schema and make sure that it matches the DataFrame that we will merge the data into:

```
df_new.printSchema()
df_new.show(truncate=False)
```

Finally, we can merge both DataFrames and display the result:

```
union_df = df.union(df_new)
union_df.show(truncate=False)
```

11. After merging both DataFrames, we can run a `select` statement to display the results of one column:

```
union_df.select("value").show()
```

12. We can also select multiple columns:

```
union_df.select("id", "value").show()
```

13. We can use `select` also in cases where we have nested structures; a way in which we can do this is as follows:

```
nested_df.select("col1.value1","col2.value2").
show(truncate=False)
```

14. We can also filter results based on a certain condition using the `filter` function:

```
union_df.filter(union_df.val1 == "c2") \
    .show(truncate=False)
```

We could have also used the `where` function, which is equivalent to `filter`.

15. In the next example, we will use the `col` and `asc` functions to reference columns and ascending order on a query that filters results based on multiple conditions. This is an `or` condition referenced using the | symbol and it's applied both in the `id` and `value` columns:

```
from pyspark.sql.functions import col, asc
filter_df = union_df.filter((col("id") == "3") |
(col("val1") == "c2")).sort(asc("id"))
display(filter_df)
```

16. In the same way, we can use the & symbol to express an and condition:

```
filter_df = union_df.filter((col("id") == "3") &
(col("val1") == "c2")).sort(asc("id"))
display(filter_df)
```

17. We can then persist the unioned DataFrame to a Parquet file by first removing the file if it exists, and then writing the file:

```
dbutils.fs.rm("/tmp/ filter_results.parquet", True)
filter_df.write.parquet("/tmp/filter_results.parquet")
```

18. If you would like to write out the DataFrame specifying to partition on a particular column, you can do it in the next way:

```
filter_df.write.\
partitionBy("id", "val1").\
parquet("/tmp/filter_results.parquet")
```

Make sure that the partition does not create a large number of partition columns otherwise the overhead of the metadata can cause processes to slow down.

19. If our data has null values, we can replace them using the fillna function. In the next example, we replace them with the "-" character:

```
non_nun_union = union_df.fillna("-")
```

20. In order to apply functions to columns, we can use the withColumn function. The following example changes the datatype of the id column to an integer:

```
non_nun_union = non_nun_union.\
withColumn("id",col("id").cast("Integer"))
```

21. We can also perform counts and aggregations on the data using the agg and countDistinct functions. We will perform an aggregation selecting the number of distinct number of values for the val column in the union_df DataFrame:

```
from pyspark.sql.functions import countDistinct
count_distinct_df = non_nun_union.select("id", "val1")\
    .groupBy("id")\
    .agg(countDistinct ("val1").alias("distinct_val1"))
display(count_distinct_df)
```

22. Finally, we can obtain describing statistics from the DataFrame:

```
count_distinct_df.describe().show()
```

23. The result of this query can then be registered as a temporary view so that we can query it using SQL commands. To do this, we can then run the following:

```
count_distinct_df.createOrReplaceTempView("count_
distinct")
```

24. After this, we can perform queries using SQL commands:

```
select_sql = spark.sql('''
  SELECT * FROM count_distinct
''')
```

25. In order to load a table registered in the Hive metastore into a DataFrame, we can use the `table` function:

```
count_distinct_df = table("count_distinct_df")
```

26. We can also cache a table that we know will be used across queries frequently. Caching is one method to speed up applications accessing the same data multiple times. We have two function calls for caching a DataFrame, `cache` and `persist`, with the difference that `cache` will cache the DataFrame into memory, and `persist` can cache in memory, disk, or off-heap memory and is equivalent to cache without an argument. To cache the previous DataFrame, we can run the following:

```
count_distinct_df.cache()
```

27. In order to clear all the cached tables on the current cache, we can do this at a global level or per table using the following commands:

```
sqlContext.clearCache()
sqlContext.cacheTable("count_distinct_df ")
sqlContext.uncacheTable("count_distinct_df ")
```

Now that we have reviewed most of the important commands of PySpark, we can make use of them when working with data in Azure Databricks.

In the next section, we will learn about a different way to manipulate data that resembles the pandas DataFrame API, and it's named Koalas.

pandas DataFrame API (Koalas)

Data scientists and data engineers that are Python users are very familiar with working with pandas DataFrames when manipulating data. `pandas` is a Python library for data manipulation and analysis but that lacks the capability to work with big data, therefore it is only suitable when working with small datasets. When we need to work with more data, the most common option is PySpark, as we have demonstrated in the previous section, which is a library with a very different syntax than pandas.

Koalas is a library that eases the learning curve from transitioning from `pandas` to working with big data in Azure Databricks. Koalas has a syntax that is very similar to the pandas API but with the functionality of PySpark.

Not all the pandas methods have been implemented and there are many small differences or subtleties that must be considered and might not be obvious. We cannot understand Koalas without understanding PySpark.

Koalas, functionality is built around the Koalas DataFrame, which is a collection of data distributed across several workers and organized in rows with named columns. This is the main difference with the panda's DataFrame, which is loaded into a single machine.

In the next section, we will find how we can migrate from working with pandas and small datasets to using the Koalas API in Azure Databricks to work with large amounts of data.

Using the Koalas API

We will describe up next how we can create different objects using the Koalas API that have their parallels in the `pandas` library. First, we will import the following packages, which are imported in the examples of the functionalities of the Koalas API and will allow us to use Koalas more flexibly:

```
import numpy as np
import pandas as pd
import databricks.koalas as ks
```

First, we will create a Koalas Series. This can be created by passing a list of values, in a very similar way in which we would create a pandas Series. A Koalas Series can also be created by a pandas Series.

A pandas Series is created in the following way:

```
pandas_series = pd.Series([1, 10, 5, np.nan, 10, 2])
```

In a very similar way, we can create a Koalas Series:

```
koala_series = ks.Series([1, 10, 5, np.nan, 10, 2])
```

As was mentioned before, we can create a Koalas Series by passing the previously created pandas Series:

```
koala_series = ks.Series(pandas_series)
```

We can also create it using the `from_pandas` function:

```
koala_series = ks.from_pandas(pandas_series)
```

One thing to bear in mind is that unlike pandas, Koalas will not guarantee maintaining the order of indices, because of the distributed nature of the Koalas API. Nevertheless, we can use `Series.sort_index()` to keep the ordered indices, as shown in the next example, where we apply `sort_index` to a Koalas Series:

```
koala_series.sort_index()
```

A Koalas DataFrame can be created in similar ways as a pandas DataFrame. For example, we can create a Koalas DataFrame by passing a NumPy array, the same way we would do for a pandas DataFrame.

To create a pandas DataFrame from a NumPy array, we can do the following:

```
pandas_df = pd.DataFrame({'val1': np.random.rand(5),
                          'val2': np.random.rand(5)})
```

Very similarly, we can do the same to create a Koalas DataFrame from a NumPy array:

```
koalas_df = ks.DataFrame({'val1': np.random.rand(5),
                          'val2': np.random.rand(5)})
```

We can also create a Koalas DataFrame by passing a pandas DataFrame as an argument to the Koalas DataFrame method:

```
koalas_df = ks.DataFrame(pandas_df)
```

Or we can use the `from_pandas` function:

```
koalas_df = ks.from_pandas(pandas_df)
```

As we would do in a pandas DataFrame, we can display the top rows using the `head` method in the Koalas DataFrame. For example, to display the first 10 rows, we can do the following:

```
koalas_df.head(10)
```

The `head` option is in fact a limit PySpark function wrapped by Koalas to make the use more similar to pandas and might also explain the possible difference in behavior between the `head` function in Koalas versus the one in pandas.

We can display a quick statistical summary of the data in our Koalas DataFrame using the `describe` function:

```
koalas_df.describe()
```

We can transpose a Koalas DataFrame by using the `transpose` function:

```
koalas_df.transpose()
```

When using this function, it is always best practice to set a maximum number of computing rows to prevent failures that might occur if the number of rows is more than the value of the `compute.max_rows` parameter, which is set to `1000` by default. Therefore, we can prevent executing expensive operations by mistake. To set this option, we can do the following, where we set the maximum computing rows to `2000`:

```
from databricks.koalas.config import set_option, get_option
ks.get_option('compute.max_rows')
ks.set_option('compute.max_rows', 2000)
```

As for selecting and accessing data stored in a Koalas DataFrame, it is also similar to a pandas DataFrame. For example, selecting a single column from a Koalas DataFrame returns a Series. In the next example, we access the `val1` column of our DataFrame:

```
koalas_df['val1']
```

We could have also accessed the column in the following way:

```
koalas_df.val1
```

We can select multiple columns from a Koalas DataFrame by passing a list of the column names we want to select:

```
koalas_df[['val1', 'val2']]
```

The `loc` and `iloc` slicing methods are also available in Koalas DataFrames. For example, to select rows from a Koalas DataFrame, we can use `loc`:

```
koalas_df.loc[1:2]
```

Slicing rows and columns is also available using the `iloc` method by referencing the index of rows and columns:

```
koalas_df.iloc[:3, 1:2]
```

In order to prevent expensive operations, in Koalas it is disabled by default to add columns coming from other DataFrames or Series. This is because these operations require joins, which are expensive operations and might lead to undesired costs.

In order to control this behavior, we can set `compute.ops_on_diff_frames` to `True` if we would like to enable adding this feature. See the available options in the docs for more detail:

```
ks.get_option('compute. on_diff_frames','true')
koalas_series = k.Series([200, 350, 150, 400, 250],
                         index=[0, 1, 2, 3, 4])
koalas_df['val3'] = koalas_series
```

After performing the column addition, we can reset this option to its default to avoid potential undesired use later on:

```
reset_option("compute.ops_on_diff_frames")
```

One of the most useful functions in pandas is the `apply` method, which is also available in Koalas DataFrames. For example, we can apply the `np.cumsum` function to sum across columns:

```
koalas_df.apply(np.cumsum)
```

We can also specify the axis in which we want to apply the function, where we can set the axis to `1` in order to work with columns. The default parameter is `0` or `index`:

```
koalas_df.apply(np.cumsum, axis=1)
```

It is best practice when using the `apply` function to specify the return datatype hint on user-defined functions. Otherwise, the Koalas API will make a trial execution of the user-defined function in order to sample the datatype, which can be computationally expensive, so it's always better to have the output datatype of the function specified.

In the next example, we see how we can define the expected return datatype in a custom function named `square_root`:

```
def square_root(x) -> ks.Series[np.float64]:
    return x ** .5
koalas_df.apply(square_root)
```

When dealing with operations that if triggered by mistake could lead to high computation costs, we can use `compute.shortcut_limit` (default = 1000), which is an option that specifies the number of rows to be computed using the pandas API. Koalas will use the pandas API to make a computation if the size of the data is below this threshold. Therefore, setting this limit too high might lead to slowing down the process execution or causing out-of-memory errors, but in turn, it saves us from triggering extensive computations.

In the next example code, we will set a higher `compute.shortcut_limit` value, which in this case has been set to `5000`:

```
ks.set_option('compute.shortcut_limit', 5000)
```

We can make line charts using the `plot.line` function:

```
koalas_df = ks.DataFrame({val1: [15, 18, 489, 675, 1776],\
                'val2': [4, 25, 181, 600, 1900]},\
              index=[1, 2, 3, 4, 5])
koalas_df.plot.line()
```

We can set the proportion of data being plotted using the `plotting.sample_ratio` option, which by default is set to `1000`. We can also do the same by setting the `plotting.max_rows` option.

The functionality to make histograms is also included in Koalas DataFrames using the `DataFrame.plot.hist` method. For example, we can create a DataFrame using random data and plotting using the previously mentioned function:

```
koalas_df = pd.DataFrame(
    np.random.randint(1, 10, 1000),\
    columns=[c1])
koalas_df[c2] = koalas_df['c1'] + np.random.randint(1, 10,
1000)
koalas_df = ks.from_pandas(koalas_df)
koalas_df.plot.hist(bins=14, alpha=0.5)
```

In the next section, we will briefly see how we can make SQL queries using the Koalas API.

Using SQL in Koalas

One of the features of Spark DataFrames is that we can query the data in many ways, one of them being using SQL queries to fetch data from tables.

This feature also extends to Koalas DataFrames, where we can make SQL queries using the Koalas API, which supports standard SQL syntax using `ks.sql`. The query returns the result as a Koalas DataFrame:

```
koalas_df = ks.DataFrame({'year': [2010, 2011, 2012, 2013,
2014],
                          'c1': [15, 24, 689, 575, 1376],
                          'c2': [4, 27, 311, 720, 1650]})
ks.sql("SELECT * FROM { koalas_df } WHERE c1 > 100")
```

Using SQL statements to query the data is a feature that acts as a bridge between data scientists used to pandas DataFrames and data engineers or business intelligence analysts who commonly use SQL to query data.

The next section will show how we can transform Koalas DataFrames into Spark DataFrames seamlessly.

Working with PySpark

You can apply several PySpark features such as the Spark UI, the history server, and so on in Koalas DataFrames, mostly because Koalas is built on top of PySpark. For example, we can easily convert Koalas DataFrames from and to PySpark DataFrames using the Koalas DataFrame `to_spark` method, which is like the `to_pandas` method. In turn, a PySpark DataFrame can be converted into a Koalas DataFrame using the PySpark DataFrame `to_koalas` method, which is an extension of the Spark DataFrame class:

```
koalas_df = ks.DataFrame({'c1': [1, 2, 3, 4, 5],
                          'c2': [10, 20, 30, 40, 50]})
spark_df = koalas_df.to_spark()
type(spark_df)
spark_df.show()
```

It is important to notice when converting a PySpark Data Frame to Koalas that this operation can cause an out-of-memory error when the default index type is a sequence. The default index type can be managed using the `compute.default_index_type` Koalas option, which by default is set to sequence. If we are dealing with indexes that are the sequence of a large dataset, we should use `distributed-sequence`:

```
from databricks.koalas import option_context
with option_context("compute.default_index_type", "distributed-sequence"):
    koalas_df = spark_df.to_koalas()
```

PySpark DataFrames do not have indices while Koalas DataFrames do. Therefore, converting a PySpark DataFrame into a Koalas DataFrame requires creating new indexes, an operation that could cause some overhead. We can avoid this issue by specifying the column that can be used as an index column, as in the next example:

```
spark_df.to_koalas(index_col='c1')
```

We can also take advantage of the PySpark caching feature, which is also available in Koalas as well. We can use this option in order to store the results of the Koalas operations and not have to run computations more than once, which can improve the performance of operations where outputs need to be accessed frequently. The `cache` option is applied to a Koalas DataFrame in the following way:

```
koalas_df.cache()
```

All of the examples presented in this section about Koalas focus on the ease of use when migrating from pandas into distributed computing in Azure Databricks and lack enough depth to explain all the available features, so it is always good to take a look at the official Koalas API documentation to keep up to date with all the available features.

In the next section of this chapter, we will learn the basics and possibilities to visualize our data in Azure Databricks using common Python libraries such as Bokeh, Matplotlib, and Plotly.

Visualizing data

We can use popular Python libraries such as Bokeh, Matplotlib, and Plotly to make visualizations in Azure Databricks. In this section, we will learn how we can use these libraries in Azure Databricks and how we can make use of notebook features to work with these visualizations.

Bokeh

Bokeh is a Python interactive data visualization library used to create beautiful and versatile graphics, dashboards, and plots.

To use Bokeh, you can install the Bokeh `PyPI` package either by installing it at the cluster level through the libraries UI and attaching it to your cluster or by installing it at the notebook level using `pip` commands.

Once we have installed the library and we can import it into our notebook, to display a Bokeh plot in Databricks, we must first create the plot and generate an HTML file embedded with the data for the plot, created, for example, by using Bokeh's `file_html` or `output_file` functions, and then pass this HTML to the Databricks `displayHTML` function.

For example, to create a simple plot using Bokeh, we can do the following. First, we import the required libraries from Bokeh:

```
from bokeh.plotting import figure
from bokeh.embed import components, file_html
from bokeh.resources import CDN
```

Then, we generate some data to visualize it. We will use a series of x and y coordinates arranged into a list to create a simple line:

```
x = [1, 2, 3, 4, 5, 6, 7, 8, 9]
y = [2, 4, 3, 4, 5, 6, 7, 9, 8]
```

Then, we can create a new object to store the line plot and add some title and axis labels to it:

```
p = figure(title="this simple line example",
           x_axis_label='x_val', y_axis_label='y_val')
```

We will also add a legend to the line and specify the line thickness:

```
p.line(x, y, legend_label="this line.", line_width=2)
```

Finally, we create the HTML file with the embedded Bokeh plot in an object:

```
html = file_html(p, CDN, "my_plot")
```

Finally, we can display it in an Azure Databricks notebook using the `displayHTML` function:

```
displayHTML(html)
```

This results in the following output:

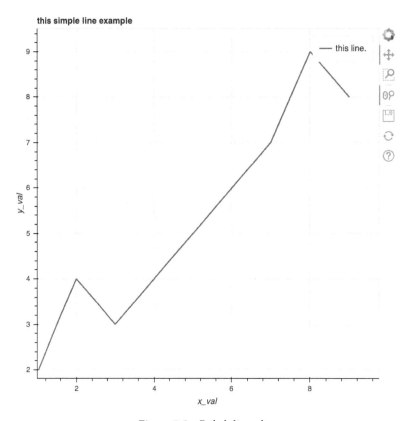

Figure 7.5 – Bokeh line plot

It is important to take into account that the maximum size for a notebook cell, both contents and output, is 16 MB, so we need to make sure that the size of the plot packed into the HTML doesn't reach this limit.

Matplotlib

The method for displaying Matplotlib figures used depends on which version of Databricks Runtime your cluster is running, but from Databricks Runtime 6.5 and above, we can directly display figures inline using the display function.

To make the case for how easy it is to work with Matplotlib, we will create a simple example displaying a scatter plot. First, we will generate fake data using numpy:

```
import numpy as np
import matplotlib.pyplot as plt
x = np.linspace(0, 2*np.pi, 50)
y = np.sin(x)
```

Then, we initialize the objects that will hold the plot and select to plot the y data over the x axis using a dashed line:

```
fig, ax = plt.subplots()
ax.plot(x, y, 'k--')
```

We can set the ticks and tick labels for both the x and y axes:

```
ax.set_xlim((0, 2*np.pi))
ax.set_xticks([0, np.pi, 2*np.pi])
ax.set_xticklabels(['0', '$\pi$','2$\pi$'])
ax.set_ylim((-1.5, 1.5))
ax.set_yticks([-1, 0, 1])
```

Finally, we can use display(fig) to display our figure inline in an Azure Databricks notebook:

```
display(fig)
```

The output is as follows:

```
Cmd 3

1   import numpy as np
2   import matplotlib.pyplot as plt
3   x = np.linspace(0, 2*np.pi, 50)
4   y = np.sin(x)
5   fig, ax = plt.subplots()
6   ax.plot(x, y, 'k--')
7   ax.set_xlim((0, 2*np.pi))
8   ax.set_xticks([0, np.pi, 2*np.pi])
9   ax.set_xticklabels(['0', '$\pi$','2$\pi$'])
10  ax.set_ylim((-1.5, 1.5))
11  ax.set_yticks([-1, 0, 1])
12  display(fig)
13
```

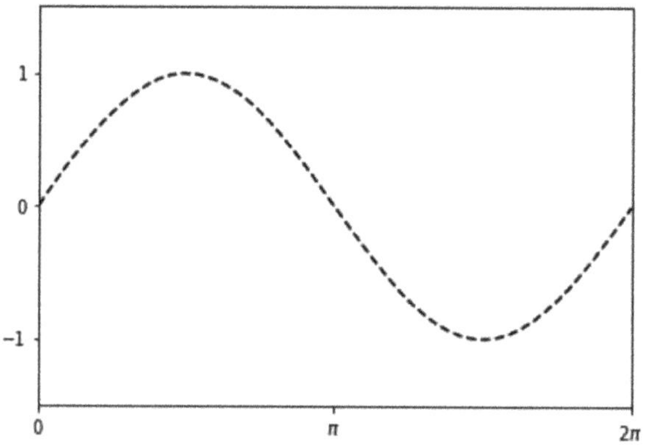

```
Command took 0.47 seconds -- by          om.ar at 09/02/2021,
```

Figure 7.6 – Matplotlib line plot

The use of Matplotlib is very similar to its use in other Python IDEs such as Jupyter notebooks, with the added difference of using the display function to visualize the plots.

Next, we will see how we can use Plotly to visualize data using the display function as well.

Plotly

Plotly is an interactive graphing library in Python that Azure Databricks supports. To use Plotly, you can either install the Plotly PyPI package and attach it to your cluster or use pip commands to install it as a notebook-scoped library.

In Databricks notebooks, it is recommended to use Plotly on its offline version and take into account that its performance may not be the best when handling large datasets. Therefore, if there are performance issues, you should try to reduce the amount of data used in the plot:

To display a Plotly plot, we must first store the resulting plot in an object and then pass this object to the displayHTML function. In order to exemplify this, let's first import Plotly and Numpy and then create some fake data to create a plot out of it:

```
from plotly.offline import plot
from plotly.graph_objs import *
import numpy as np
Then, we can create some random data to plot:
x = np.random.randn(1000)
y = np.random.randn(1000)
```

After this, we will create the plot but instead of calling the plot function to display the visualization, we will store the plot in an object, making sure that we pass the output_type keyword argument set to div:

```
p = plot(
    [
        Histogram2dContour(x=x, y=y,
contours=Contours(coloring='heatmap')),
        Scatter(x=x, y=y, mode='markers',
marker=Marker(color='white', size=3, opacity=0.3))
    ],
    output_type='div'
)
```

Finally, we can display the plot by calling the displayHTML function:

```
displayHTML(p)
```

The output is as follows:

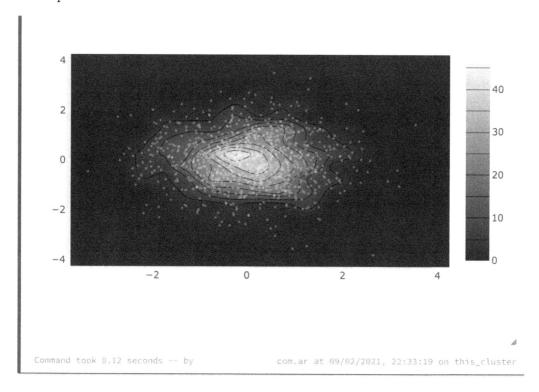

Figure 7.7 – Plotly scatter plot

As we have seen in this section, using common libraries to visualize data is really easy once we understand the minor differences between working locally and using these libraries in Azure Databricks.

Summary

In this chapter, we have turned our focus to deepening our knowledge on how we can manipulate data using the PySpark API and using its core features. We have also gone through the steps required to install and use Python libraries at different instance levels and how we can use them to visualize our data using the display function.

Finally, we went through the basics of the Koalas API, which makes it easier to migrate from working with pandas to working with big data in Azure Databricks.

In the next chapter, will learn how to use Azure Databricks to run machine learning experiments, train models, and make inferences on new data using libraries such as XGBoost, sklearn, and Spark's MlLib.

8
Databricks Runtime for Machine Learning

This chapter will be a deep dive into the development of classic machine learning algorithms to train and deploy models based on tabular data, exploring libraries and algorithms as well. The examples will be focused on the particularities and advantages of using Azure Databricks Runtime for Machine Learning (Databricks Runtime ML).

In this chapter we will explore the following concepts, which are focused on how we can extract and improve the features available in our data to train our machine learning and deep learning models. The topics that we will cover are listed here:

- Loading data
- Feature engineering
- Time-series data sources
- Handling missing values
- Extracting features from text
- Training machine learning models on tabular data

In the following sections, we will discuss the necessary libraries needed to perform the operations introduced, as well as providing some context on how best practices and some core concepts are related to them.

Without further ado, let's start working with machine and deep learning models in Azure Databricks.

Loading data

Comma-separated values (**CSV**) are the most widely used format for tabular data in machine learning applications. As the name suggests, it stores data arranged in the form of rows, separated by commas or tabs.

This section covers information about loading data specifically for machine learning and deep learning applications. Although we can consider these concepts covered in the previous chapters and sections, we will reinforce concepts around how we can read tabular data directly into Azure Databricks and which are the best practices to do this.

Reading data from DBFS

When training machine learning algorithms in a distributed computing environment such as Azure Databricks, the need to have shared storage becomes important, especially when working with distributed deep learning applications. Azure **Databricks File System** (**DBFS**) allows efficient access to data for any cluster using Spark and local file **application programming interfaces** (**APIs**):

In Azure Databricks, we have an option to choose a Machine Learning Runtime that provides a high-performance **Filesystem in Userspace** (**FUSE**) mount, which is a virtual filesystem that can be accessed in the /dbfs location by all cluster nodes. This allows for the process that is running in cluster nodes to read to the same location using the distributed storage system with local file APIs.

- For example, when we write a file to the DBFS location, as illustrated in the following code snippet, we are using the shared filesystem across all cluster nodes:

```
with open("/dbfs/tmp/test_dbfs.txt", 'w') as f:
    f.write("This is\n")
    f.write("in the shared\n")
    f.write("file system.\n")
```

- In the same way, we can read a file on the distributed filesystem using local file APIs, as illustrated in the following code snippet:

```
with open("/dbfs/tmp/test_dbfs.txt", "r") as f_read:
    for line in f_read:
        print(line)
```

- This way, we can display the output to the console, as shown in the following screenshot:

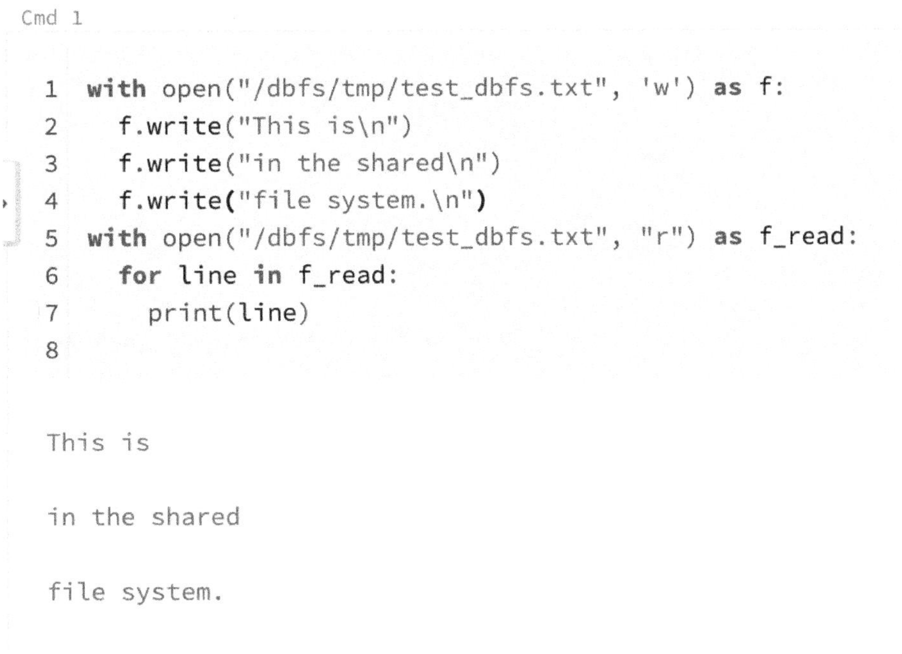

```
Cmd 1

1  with open("/dbfs/tmp/test_dbfs.txt", 'w') as f:
2    f.write("This is\n")
3    f.write("in the shared\n")
4    f.write("file system.\n")
5  with open("/dbfs/tmp/test_dbfs.txt", "r") as f_read:
6    for line in f_read:
7      print(line)
8


This is

in the shared

file system.
```

Figure 8.1 – Output of reading a file in the shared filesystem

When using Azure Databricks Runtime ML, we have the option to use the dbfs:/ml folder, which is a special folder that has been optimized to offer high-performance **input/ output (I/O)** for deep learning operations. This is especially handy as a workaround to the limitation of supporting files that are smaller than 2 **gigabytes (GB)** in Azure Databricks, so it is recommended to save data in this folder.

We can load data for training and inference of machine and deep learning applications by reading from tables or local CSV files, as we will see in the next section. Once the data has been loaded, we have plenty of options to then convert it to Spark DataFrames, pandas DataFrames, or NumPy arrays.

In the next section, we will again go through how we can read CSV files to work with distributed deep learning applications in Azure Databricks.

Reading CSV files

One of the most frequent formats for machine learning data is CSV, and it's regularly used to hold tabular data. Here, we use the term tabular data to describe data structured as rows representing observations, which in turn are described using variables or attributes in the form of columns.

As we have seen in previous chapters, we have a number of ways to load a CSV file into Azure Databricks. The first option would be to import the data into a Spark DataFrame using the PySpark API to parse the data. The following piece of code will read the file stored in the specified `databricks` dataset folder in DBFS into a Spark DataFrame inferring the underlying data types in the file:

```
diamonds = spark.read.format('csv').options(header='true',
inferSchema='true').load('/databricks-datasets/Rdatasets/data-
001/csv/ggplot2/diamonds.csv')
```

In Azure Databricks, we have also the option of using **Structured Query Language (SQL)** commands to import data from a CSV file into a temporary view that we can then query. We can do this by using `%sql` magic in a cell and loading the data in the CSV file into a temporary view, as follows:

```
%sql
CREATE TEMPORARY VIEW diamonds
USING CSV
OPTIONS (path "/databricks-datasets/Rdatasets/data-001/csv/
ggplot2/diamonds.csv", header "true", mode "FAILFAST")
```

Pay attention to the fact that here, we pass the `FAILFAST` mode to stop the parsing of the file, causing an exception if faulty lines are found in it.

Once the data has been imported into the view, we can use SQL commands to perform queries on the temporary view, as follows:

```
%sql
SELECT * FROM diamonds
```

This will show us the temporary view that we just created, as illustrated in the following screenshot:

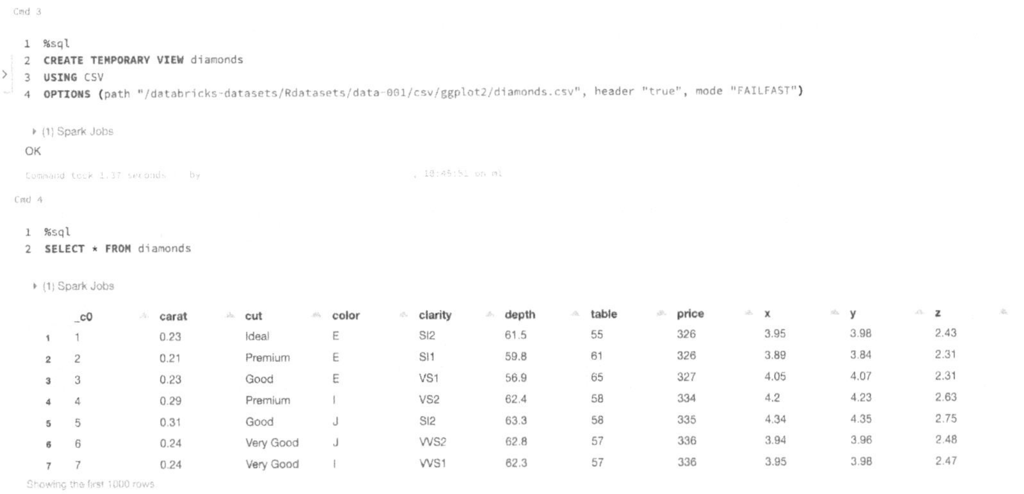

Figure 8.2 – Using SQL to read from a CSV file

If we read a CSV file specifying the schema, we may encounter some issues if the specified schema differs from what we have specified—for example, we may have a string value in a column that was specified as an integer data type. This could lead to results that differ considerably from the actual data. Therefore, it is always good practice to verify the correctness of the data.

We can specify the behavior of the CSV parser by selecting one of the available modes in which the parser runs. The available options for this mode are outlined here:

- **PERMISSIVE**: This is the default mode. Here, null values are inserted in fields that have not been parsed correctly. This allows us to inspect rows that have not been parsed correctly.

- **DROPMALFORMED**: This mode drops lines that hold values that could not be parsed correctly.

- **FAILFAST**: This mode aborts the reading of the file if any malformed data is found.

To set the mode, use the mode option, as illustrated in the following code snippet:

```
drop_wrong_schema = sqlContext.read.format("csv").
option("mode", "FAILFAST")
```

In the next example, we will read a CSV file, specifying a known schema of the data with the parser mode set to drop malformed data if its type is not as specified. As discussed before, specifying the schema is always a good practice. If the schema of the data is already known, it is better to avoid schema inference. Here, we will read a file with columns for ID, name, and last name, the first as an integer and the rest as strings:

```python
from pyspark.sql.types import StructType, StructField
from pyspark.sql.types import DoubleType, IntegerType,
StringType
data_schema = StructType([
    StructField("id", IntegerType()),
    StructField("name", StringType()),
    StructField("lastname", StringType())
])
(sqlContext
    .read
    .format("com.databricks.spark.csv")
    .schema(data_schema)
    .option("header", "true")
    .option("mode", "DROPMALFORMED")
    .load("some_input_file.csv"))
```

Now that we have gone through how to read tabular data from CSV files, we can use the data to extract features from it. These features will allow us to accurately model our data and have more effective machine learning and deep learning models. In the next section, we will learn how to perform feature engineering in Azure Databricks.

Feature engineering

Machine learning models are trained using input data to later provide as an outcome a prediction on unseen data. This input data is regularly composed of features that usually come in the form of structured columns. The algorithms use this data in order to infer patterns that may be used to infer the result. Here, the need for feature engineering arises with two main goals, as follows:

- Refactoring the input data to make it compatible with the machine learning algorithm we have selected for the task. For example, we need to encode categorical values if these are not supported by the algorithm we choose for our model.

- Improving the predictions produced by the models according to the performance metric we have selected for the problem at hand.

With feature engineering, we extract relevant features from the raw input data to be able to accurately represent it according to the way in which the problem to be solved has been modeled, resulting in an improvement in the performance model on novel data. It's good to keep in mind that the features used will influence the performance of the model more than everything else.

Feature engineering can be considered the process of transforming the input data into something that is easier to interpret for the machine learning algorithm by making key features or patterns more transparent, although we can sometimes generate new features to make the data visualization more interpretable or the pattern inference clearer for the model algorithm.

There are multiple strategies applied in feature engineering. Some of them are listed here:

- Missing data imputation and management
- Outlier handling
- Binning
- Logarithmic transformations
- One-hot encoding
- Normalization
- Grouping and aggregation operations
- Feature splitting
- Scaling

Some of these strategies have their application in just some algorithms or datasets, while others can be beneficial in all cases.

It is a common belief to think that the better the features you model, the better the performance you will achieve. While this is true in certain circumstances, it can also be misleading.

The performance of the model has many factors, with different degrees of importance. Among those factors are the selection of the algorithm, the available data, and the features extracted. You can also add to those factors the way in which the problem is modeled, and the objective metrics used to estimate performance. Nevertheless, great features are needed to effectively describe the structures inherent in your data and will make the job of the algorithm easier.

When working with large amounts of data, tools such as Spark SQL and MLlib can be very effective when used in feature engineering. Some third-party libraries are also included in Databricks Runtime ML, such as scikit-learn, which provides useful methods to extract features from data.

This section does not intend to go too deep into the mechanics of each algorithm, although some concepts will be reinforced to understand how they are applied in Azure Databricks.

Tokenizer

Tokenization can be generally described as a process in which we transform a string of input characters into an array formed by the individual sections that form it. These sections are called tokens and are generally individual words, but can be a certain number of characters, or a combination of words called n-grams. We will use these resulting tokens later on for some other form of processing to extract or transform features. In this sense, the tokenization process can be considered a task of feature engineering. The tokens are identified based on the specific modes that are passed to the parser.

In PySpark, we can use the simple Tokenizer class to tokenize a sequence of string inputs. The following code example shows how we can split sentences into sequences of words:

```
from pyspark.ml.feature import Tokenizer
sentenceDataFrame = sqlContext.createDataFrame([
    (0, "Spark is great for Data Science"),
    (0, "Also for data engineering"),
    (1, "Logistic regression models are neat")
], ["label", "sentence"])
tokenizer = Tokenizer(inputCol="sentence", outputCol="words")
wordsDataFrame = tokenizer.transform(sentenceDataFrame)
for words_label in wordsDataFrame.select("words", "label").
take(3):
    print(words_label)
```

Here, we can see the encoded tokens:

Cmd 5

```
1   from pyspark.ml.feature import Tokenizer
2   sentenceDataFrame = sqlContext.createDataFrame([
3     (0, "Spark is great for Data Science"),
4     (0, "Also for data engineering"),
5     (1, "Logistic regression models are neat")
6   ], ["label", "sentence"])
7   tokenizer = Tokenizer(inputCol="sentence", outputCol="words")
8   wordsDataFrame = tokenizer.transform(sentenceDataFrame)
9   for words_label in wordsDataFrame.select("words", "label").take(3):
10    print(words_label)
```

▸ (2) Spark Jobs

▸ 🖻 sentenceDataFrame: pyspark.sql.dataframe.DataFrame = [label: long, sentence: string]

▸ 🖻 wordsDataFrame: pyspark.sql.dataframe.DataFrame = [label: long, sentence: string ... 1 more fields]

Row(words=['spark', 'is', 'great', 'for', 'data', 'science'], label=0)
Row(words=['also', 'for', 'data', 'engineering'], label=0)
Row(words=['logistic', 'regression', 'models', 'are', 'neat'], label=1)

Figure 8.3 – Output of the word tokenization

We can also use the more advanced RegexTokenizer tokenizer to specify token boundaries using regular expressions.

Binarizer

Binarization is a process in which you establish a numerical threshold on your features in order to transform them into binary features. Binarization is really useful to preprocess input data that holds continuous numerical features if you can assume that our data has a probabilistic normal distribution. This process makes the data easier to handle for the algorithms used in machine learning and deep learning because they describe the data into a more defined structure.

In PySpark, we can use the simple Binarizer class, which allows us to binarize continuous numerical features. Besides the common parameters of inputCol and outputCol, the simple Binarizer class has a parameter threshold that is used to establish the threshold to binarize the continuous numerical features. The features greater than this threshold are binarized to 1.0, and the ones equal or less than this threshold are binarized to 0.0. The following code example shows how we can use the Binarizer class to binarize numerical features. First, we import the Binarizer class and create an example DataFrame with label and feature columns:

```
from pyspark.ml.feature import Binarizer
continuousDataFrame = sqlContext.createDataFrame([
```

```
   (0, 0.345),
   (1, 0.826),
   (2, 0.142)
], ["label", "feature"])
```

Next, we instantiate the `Binarizer` class, specifying a threshold and the input and output columns. Then, we transform the example DataFrame and store and display the results. The code to do this is illustrated in the following snippet:

```
binarizer = Binarizer(threshold=0.5, inputCol="feature",
                      outputCol="binarized_feature")
binarizedDataFrame = binarizer.transform(continuousDataFrame)
binarizedFeatures = binarizedDataFrame.select("binarized_
feature")
for binarized_feature, in binarizedFeatures.collect():
   print(binarized_feature)
```

We should see printed the binarized features of the transformed DataFrame according to the threshold that we have established. Next, we will learn how to apply a polynomial expansion to express features in a higher dimensional space.

Polynomial expansion

In mathematics, a polynomial expansion is a mathematical operation used to express a feature as a product of other higher-level features. Therefore, it can be used to expand features passed to express them as a set of features in a higher dimension.

A polynomial expansion can be considered a process of expanding your features into a polynomial space, in order to try to find patterns in that expanded space that otherwise might be difficult to find in a lower-dimension space. This is common when dealing with algorithms that make simple assumptions, such as linear regression, that might otherwise be unable to capture relevant patterns. This polynomial space in which we transform our features is formed by an n-degree combination of original dimensions. The PySpark `PolynomialExpansion` class provides us with this functionality. The following code example shows how you can use it to expand your features into a 3-degree polynomial space:

```
from pyspark.ml.feature import PolynomialExpansion
from pyspark.ml.linalg import Vectors
df = spark.createDataFrame([
    (Vectors.dense([2.0, 1.0]),),
```

```
    (Vectors.dense([0.0, 0.0]),),
    (Vectors.dense([3.0, -1.0]),)
], ["features"])
polyExpansion = PolynomialExpansion(degree=3,
inputCol="features", outputCol="polyFeatures")
polyDF = polyExpansion.transform(df)
polyDF.show(truncate=False)
```

Here, we can see the variables that were created in the polynomial expansion:

> Cmd 7

```
1    from pyspark.ml.feature import PolynomialExpansion
2    from pyspark.ml.linalg import Vectors
3
4    df = spark.createDataFrame([
5        (Vectors.dense([2.0, 1.0]),),
6        (Vectors.dense([0.0, 0.0]),),
7        (Vectors.dense([3.0, -1.0]),)
8    ], ["features"])
9
10   polyExpansion = PolynomialExpansion(degree=3, inputCol="features", outputCol="polyFeatures")
11   polyDF = polyExpansion.transform(df)
12
13   polyDF.show(truncate=False)
```

▸ (2) Spark Jobs
  ▸ 🖾 df: pyspark.sql.dataframe.DataFrame = [features: udt]
  ▸ 🖾 polyDF: pyspark.sql.dataframe.DataFrame = [features: udt, polyFeatures: udt]

```
+----------+------------------------------------------+
|features  |polyFeatures                              |
+----------+------------------------------------------+
|[2.0,1.0] |[2.0,4.0,8.0,1.0,2.0,4.0,1.0,2.0,1.0]     |
|[0.0,0.0] |[0.0,0.0,0.0,0.0,0.0,0.0,0.0,0.0,0.0]     |
|[3.0,-1.0]|[3.0,9.0,27.0,-1.0,-3.0,-9.0,1.0,3.0,-1.0]|
+----------+------------------------------------------+
```

Figure 8.4 – Output features of the polynomial expansion

Having our features transformed into a higher-dimension space is a great tool to use in feature engineering. It allows us to describe data in different terms and increase the sensitivity of the machine learning algorithm to more complex patterns that might be hidden in the data.

StringIndexer

The PySpark `StringIndexer` class encodes a string column of labels into a column of labeled indices. These indices are in the range from `0` to `numLabels`, ordered by label frequencies. Therefore, the most frequent label will be indexed as `0`. If the input column is numeric, the PySpark `StringIndexer` class will cast it to a string and index the string values.

The `StringIndexer` class takes an input column name and an output column name. Here, we will show how we can use it to index a column named `cluster`:

```python
from pyspark.ml.feature import StringIndexer
df = sqlContext.createDataFrame(
    [(0, "a"), (1, "b"), (2, "c"), (3, "a"),
     (4, "a"), (5, "c")],
    ["id", "cluster"])
indexer = StringIndexer(inputCol="cluster",
                        outputCol="categoryIndex")
indexed = indexer.fit(df).transform(df)
indexed.show()
```

In the following screenshot, we can see that the strings were correctly indexed into a DataFrame of variables:

```
Cmd 8
>
  1  from pyspark.ml.feature import StringIndexer
  2  df = sqlContext.createDataFrame(
  3      [(0, "a"), (1, "b"), (2, "c"), (3, "a"), (4, "a"), (5, "c")],
  4      ["id", "cluster"])
  5  indexer = StringIndexer(inputCol="cluster", outputCol="categoryIndex")
  6  indexed = indexer.fit(df).transform(df)
  7  indexed.show()

  ▶ (4) Spark Jobs
  ▶ ▦ df: pyspark.sql.dataframe.DataFrame = [id: long, cluster: string]
  ▶ ▦ indexed: pyspark.sql.dataframe.DataFrame = [id: long, cluster: string ... 1 more fields]

+---+-------+-------------+
| id|cluster|categoryIndex|
+---+-------+-------------+
|  0|      a|          0.0|
|  1|      b|          2.0|
|  2|      c|          1.0|
|  3|      a|          0.0|
|  4|      a|          0.0|
|  5|      c|          1.0|
+---+-------+-------------+
```

Figure 8.5 – Output transformation of the StringIndexer class

In this example, the a cluster gets index 0 because it is the most frequent, followed by c with index 1 and b with index 2.

One-hot encoding

One-hot encoding is a transformation used in feature engineering to create a group of binary values (often called dummy variables) that are mostly used to represent categorical variables with multiple possible values as variables with just a single high value (1) and all others a low value (0). Each of these Boolean features represents a single possible value in the original categorical variable.

One-hot encoding is sometimes used either for visualization, model efficiency, or to prepare your data according to the requirements of the training algorithms. For this, we construct different features by encoding the categorical variables. Instead of a single feature column with several levels, we split this column into Boolean features for each level, where the only accepted values are either 1 or 0.

One-hot encoding is a technique that is especially popular with neural networks and other algorithms in deep learning, to encode categorical features allowing algorithms that expect continuous features, such as logistic regression, to use these categorical features. In PySpark and Azure Databricks, we can use the OneHotEncoder class to map a column of label indices to an encoded column of binary vectors, with at most a single possible value. Here, we use the OneHotEncoder class to convert categorical features into numerical ones:

```python
from pyspark.ml.feature import OneHotEncoder
df = spark.createDataFrame([
        (0.0, 1.0),
        (1.0, 0.0),
        (2.0, 1.0),
        (0.0, 2.0),
        (0.0, 1.0),
        (2.0, 0.0)
], ["clusterV1", "clusterV2"])
encoder = OneHotEncoder(inputCols=["clusterV1",
                                   "clusterV2"],
                        outputCols=["catV1", "vatV2"])
model = encoder.fit(df)
encoded = model.transform(df)
encoded.show()
```

Here, we see the encoded variables that we obtained:

> Cmd 9

```
1   from pyspark.ml.feature import OneHotEncoder
2   df = spark.createDataFrame([
3       (0.0, 1.0),
4       (1.0, 0.0),
5       (2.0, 1.0),
6       (0.0, 2.0),
7       (0.0, 1.0),
8       (2.0, 0.0)
9   ], ["clusterV1", "clusterV2"])
10  encoder = OneHotEncoder(inputCols=["clusterV1", "clusterV2"],
11                          outputCols=["catV1", "vatV2"])
12  model = encoder.fit(df)
13  encoded = model.transform(df)
14  encoded.show()
```

▶ (3) Spark Jobs

▶ 🖻 df: pyspark.sql.dataframe.DataFrame = [clusterV1: double, clusterV2: double]

▶ 🖻 encoded: pyspark.sql.dataframe.DataFrame = [clusterV1: double, clusterV2: double ... 2 more fields]

```
+---------+---------+-------------+-------------+
|clusterV1|clusterV2|        catV1|        vatV2|
+---------+---------+-------------+-------------+
|      0.0|      1.0|(2,[0],[1.0])|(2,[1],[1.0])|
|      1.0|      0.0|(2,[1],[1.0])|(2,[0],[1.0])|
|      2.0|      1.0|    (2,[],[])|(2,[1],[1.0])|
|      0.0|      2.0|(2,[0],[1.0])|    (2,[],[])|
|      0.0|      1.0|(2,[0],[1.0])|(2,[1],[1.0])|
|      2.0|      0.0|    (2,[],[])|(2,[0],[1.0])|
+---------+---------+-------------+-------------+
```

Figure 8.6 – Output of the one-hot encoding

In this way, we can use one-hot encoding to transform categorical features in one or many columns of a Spark DataFrame into numerical features that can be later transformed or directly passed to our model to train and infer.

VectorIndexer

In Azure Databricks, we can use the `VectorIndexer` class to index categorical features as an alternative to one-hot encoding. This class is used in data frames that contain columns of type Vector. It works by deciding which features are categorical and converts those features' original values into categorical indexes. The steps that it follows are outlined next:

1. It takes as an input a column of type Vector and a `maxCategories` parameter, which—as the name suggests—specifies the threshold of a maximum number of categories for a single variable.

2. It decides which features are categorical, based on the number of distinct values that they hold. The features with at most `maxCategories` will be marked as categorical.

3. It computes 0-based category indices for each feature that was declared as categorical.

4. Finally, it indexes the categorical features by transforming the original feature values into indices.

This indexing of categorical features allows us to overcome the limitations of certain algorithms such as Decision Trees and Tree Ensembles regarding handling categorical features to improve performance or to simply be able to use algorithms that allow just numerical features.

In the following code example, we will read an example dataset of labeled points and use the PySpark `VectorIndexer` class to decide which of the features that compose the dataset should be treated as categorical variables. We will later transform the values of those categorical features into indices:

```
from pyspark.ml.feature import VectorIndexer
from pyspark.mllib.util import MLUtils
data = MLUtils.loadLibSVMFile(sc, "data/mllib/sample_libsvm_
data.txt").toDF()
indexer = VectorIndexer(inputCol="features",
                        outputCol="indexed",
                        maxCategories=10)
indexerModel = indexer.fit(data)
# Create new column "indexed" with categorical values
transformed to indices
indexedData = indexerModel.transform(data)
```

This categorical data transformed into continuous can be then be passed to algorithms such as `DecisionTreeRegressor` in order to be able to handle categorical features.

Normalizer

Another common feature engineering method is to bring the data into a given interval. A first reason to do this is to limit the computations on a fixed range of values to prevent numerical inaccuracies and limit the computational power required that might arise when dealing with numbers that are either too big or too small. A second reason to normalize the numerical values of a feature is that some machine learning algorithms will handle data better when it has been normalized. There are several approaches that we can take in order to normalize our data.

These different approaches are due to some machine learning algorithms requiring different normalization strategies in order to perform efficiently. For example, in the case of **k-nearest neighbors (KNN)**, the range of values in a particular feature impacts the weight of that feature in the model. Therefore, the bigger the values, the more importance the feature will have. In the case of the neural networks, this normalization might not impact the final performance results per se, but it will speed up the training and will avoid the exploiting and vanishing gradient problem that is caused because of the propagation of values that are either too big or too small to subsequent layers in the model. On the other hand, the decision tree-based machine learning algorithms neither benefit nor get hurt by the normalization.

The right normalization strategy will depend on the problem and selected algorithm, rather than some general statistical or computational considerations, and the right selection will always be closely tied to the domain knowledge.

In PySpark, the `Normalizer` class is a Transformer that transforms a dataset of `Vector` rows, normalizing each Vector to have the norm of the vector transformed into a unit. It takes as a parameter the value p, which specifies the p-norm used for normalization. This p value is set to 2 by default, which is equal to saying that we are transforming into the Euclidean norm.

The following code example demonstrates how to load a dataset in `libsvm` format and then normalize each row to have a unit norm:

```
from pyspark.mllib.util import MLUtils
from pyspark.ml.feature import Normalizer
data = MLUtils.loadLibSVMFile(sc, "data/mllib/sample_libsvm_
data.txt")
dataFrame = sqlContext.createDataFrame(data)
```

```
# Normalize each Vector using $L^1$ norm.
normalizer = Normalizer(inputCol="features",
                        outputCol="normFeatures", p=1.0)
l1NormData = normalizer.transform(dataFrame)
# Normalize each Vector using $L^\infty$ norm.
lInfNormData = normalizer.transform(dataFrame, {normalizer.p:
float("inf")})
```

This normalization process is widely used in feature engineering and will help us to standardize the data and improve the behavior of learning algorithms.

StandardScaler

In many areas of nature, things tend to be governed by normal or Gaussian probabilistic distribution. Normalization is what we call in feature engineering the process of subtracting the mean value of a feature, which allows us to center our feature around 0. Then, we divide this by the standard deviation, which will tell us about the spread of our feature around 0. Finally, we will obtain a variable that is centered at 0, with a range of values between -1 and 1.

In Azure Databricks, we can use the PySpark `StandardScaler` class to transform a dataset of Vector rows into normalized features that have unit standard deviation and or 0 mean. It takes the following parameters:

- `withStd`: This parameter is set to true by default. It scales the data to unit standard deviation, which means that we divide by the standard deviation.

- `withMean`: This parameter is set to false by default. It will center the data with mean equal to 0 before scaling. It creates a dense output, so this will not work on sparse input and will raise an exception.

The PySpark `StandardScaler` class is a transformation that can be fit on a dataset to produce a `StandardScalerModel`; this latter model can, later on, be used to compute summary statistics on the input data. The `StandardScalerModel` can then be transformed into a column Vector in a dataset with unit standard deviation and/or 0 mean features.

Taking into account that if the standard deviation of a feature is already 0, `StandardScalerModel` will return as default 0.0 as the value in the Vector for that feature.

In the following code example, we will show how we can load a dataset that is in `libsvm` format and then normalize each feature to have a unit standard deviation.

1. First, we will make the necessary imports and read the dataset, which is an example dataset in the `mllib` folder:

```
from pyspark.mllib.util import MLUtils
from pyspark.ml.feature import StandardScaler
data = MLUtils.loadLibSVMFile(sc, "data/mllib/sample_
libsvm_data.txt")
dataFrame = sqlContext.createDataFrame(data)
scaler = StandardScaler(inputCol="features",
                        outputCol="scaledFeatures",
                        withStd=True, withMean=False)
```

2. Then, we can compute summary statistics by fitting the `StandardScaler` class to the DataFrame created from the input data, as follows:

```
scalerModel = scaler.fit(dataFrame)
```

3. Finally, we can normalize each feature to have a unit standard deviation, as follows:

```
scaledData = scalerModel.transform(dataFrame)
```

Normalization is widely used in data science. It helps us put all continuous numerical values on the same scale and can help us overcome numerous problems that arise when the continuous variables used in our learning algorithms are not bounded to fixed ranges and are not being centered.

Bucketizer

The PySpark `Bucketizer` class will transform a column of continuous features into a column of feature buckets, where these buckets are specified by users using the splits parameter:

The splits parameter is used for mapping continuous features into buckets. Specifying $n+1$ splits will yield n buckets. A bucket which is defined by splits x and y will hold values in the range $[x,y)$ except for the last bucket, which will also include values in the range of y. These splits have to be strictly increasing. Keep in mind that otherwise specified, values outside the splits specified will be treated as errors. Examples of splits can be seen in the following code snippet:

```
splits = Array(Double.NegativeInfinity, 0.0, 1.0,
                Double.PositiveInfinity)
splits = Array(0.0, 1.0, 2.0)
```

The split values must be explicitly provided to cover all *Double* values.

Take into account that if you don't know which are the upper bounds and lower bounds of the targeted column, it is best practice to add `Double.NegativeInfinity` and `Double.PositiveInfinity` as the bounds of your splits to prevent any potential error that might arise because of an "`out of Bucketizer bounds`" exception. Note also that these splits have to be specified in strictly increasing order.

In the following example, we will show how you can bucketize a column of `Double` values into another index-wised column using the PySpark `Bucketizer` class:

1. First, we will make the necessary imports and define the DataFrame to be bucketized, as follows:

    ```
    from pyspark.ml.feature import Bucketizer
    splits = [-float("inf"), -0.5, 0.0, 0.5, float("inf")]
    data = [(-0.5,), (-0.3,), (0.0,), (0.2,)]
    dataFrame = sqlContext.createDataFrame(data,
                                    ["features"])
    bucketizer = Bucketizer(splits=splits,
                            inputCol="features",
                            outputCol="bucketedFeatures")
    ```

2. Then, we can transform the original data into its bucket index, as follows:

    ```
    bucketedData = bucketizer.transform(dataFrame)
    display(bucketedData)
    ```

3.   This is the obtained bucketized data:

```
Cmd 11
1  from pyspark.ml.feature import Bucketizer
2  splits = [-float("inf"), -0.5, 0.0, 0.5, float("inf")]
3  data = [(-0.5,), (-0.3,), (0.0,), (0.2,)]
4  dataFrame = sqlContext.createDataFrame(data, ["features"])
5  bucketizer = Bucketizer(splits=splits, inputCol="features", outputCol="bucketedFeatures")
6  #Then we can transform original data into its bucket index.
7  bucketedData = bucketizer.transform(dataFrame)
8  display(bucketedData)
```

▸ (2) Spark Jobs

▸ 🔲 dataFrame: pyspark.sql.dataframe.DataFrame = [features: double]

▸ 🔲 bucketedData: pyspark.sql.dataframe.DataFrame = [features: double, bucketedFeatures: double]

	features ▲	bucketedFeatures ▲
1	-0.5	1
2	-0.3	1
3	0	2
4	0.2	2

Showing all 4 rows.

Figure 8.7 – Bucketizer output features

4.   In PySpark, we can also use the QuantileDiscretizer class over Bucketizer where QuantileDiscretizer is an estimator that is able to handle **Not a Number** (NaN) values, and this is where the difference lies between the two of them, because the Bucketizer class is a transformer that raises an error if the input data holds NaN values. The following code snippet provides an example of a situation in which both classes yield similar outputs:

```
from pyspark.ml.feature import QuantileDiscretizer
from pyspark.ml.feature import Bucketizer
data = [(0, 18.0), (1, 19.0), (2, 8.0), (3, 5.0), (4, 2.2)]
df = spark.createDataFrame(data, ["id", "hour"])
result_discretizer = QuantileDiscretizer(numBuckets=3,
inputCol="hour",
outputCol="result").fit(df).transform(df)
result_discretizer.show()
splits = [-float("inf"),3, 10, float("inf")]
```

```
result_bucketizer = Bucketizer(splits=splits,
inputCol="hour",
outputCol="result").transform(df)
result_bucketizer.show()
```

Here the `QuantileDiscretizer` class will determine the bucket splits based on the data, while the `Bucketizer` class will arrange the data into buckets based on what you have specified via splits.

Therefore, it is better to use `Bucketizer` when you already have knowledge of which buckets you expect, while the `QuantileDiscretizer` class will estimate the splits for you.

In the preceding example, the outputs of both processes are similar because of the input data and the splits selected. Otherwise, results may vary significantly in different scenarios.

Element-wise product

An element-wise product is a mathematical operation very commonly used in feature engineering when working with data that is arranged as a matrix. The PySpark `ElementwiseProduct` class will multiply each input vector by a previously specified weight vector. To do this, it will use an element-wise multiplication that will scale each column of the input vector by a scalar multiplier defined as the weight.

The `ElementwiseProduct` class takes `scalingVec` as the main parameter, which is the transforming vector of this operation.

The following example will exemplify how we can transform vectors using the `ElementwiseProduct` class and a transforming vector value:

```
from pyspark.mllib.linalg.distributed import RowMatrix
v1 = sc.parallelize([[2.0, 2.0, 2.0], [3.0, 3.0, 3.0]])
mat = RowMatrix(v1)
v2 = Vectors.dense([0.0, 1.0, 2.0])
transformer = ElementwiseProduct(v2)
transformedData = transformer.transform(mat.rows)
print transformedData.collect()
```

This operation can then be understood as the Hadamard product, which is also known as the element-wise product between the input array and the weight vector, which in turn will give as a result another vector.

Time-series data sources

In data science and engineering, one of the most common challenges is temporal data manipulation. Datasets that hold geospatial or transactional data, which mostly lie in the financial and economics area of an application, are some of the most common examples of data that is indexed by a timestamp. Working in areas such as finance, fraud, or even socio-economic temporal data ultimately leads to the need to join, aggregate, and visualize data points.

This temporal data regularly comes in datetime formats that might vary not only in the format itself but in the information that it holds. One of the examples of this is the difference between the DD/MM/YYYY and MM/DD/YYYY format. Misunderstanding these different datetime formats could lead to failures or wrongly formed results if the formats used don't match up. Moreover, this data doesn't come in numerical format, which—as we have seen in previous sections of the chapter—can lead to several impediments that need to be overcome, one of them being the fact that this data cannot be interpreted easily by most of the learning algorithms used in machine and deep learning.

This is where feature engineering comes into play, providing us with the tools to transform and create new features from this data. An example of this could be to reorganize the data in numerical features such as day, month, and year, and even manipulate features using techniques as dynamic time warping to compare time series of different lengths, as could be the case when comparing the sales of the months of February and March, which have 28 and 31 days respectively.

In Azure Databricks, we have several functionalities in place that allow us to perform operations such as joins, aggregations, and windowing of time series, with the added benefit of doing this processing in parallel. Moreover, the Koalas API allows us to work with this data using a Pandas-like syntax that makes the transition from experiment to production much easier to handle.

In this example, we will work with financial data to illustrate how we can manipulate temporal datasets in Azure Databricks. The data that we will use is an example that is based on the stock market and holds trade information for different companies that are being traded. You can find more information about this kind of data at this link: https://www.tickdata.com/product/nbbo/:

1. We can get example data by running the following code in an Azure notebook cell. This will download the data that we will read afterward:

```
%sh
wget https://pages.databricks.com/rs/094-YMS-629/images/
ASOF_Quotes.csv ;
```

```
wget https://pages.databricks.com/rs/094-YMS-629/images/
ASOF_Trades.csv ;
```

Here, we have downloaded two kinds of datasets, one for trades and one for offers. We will merge this data afterward over the time column to compare the trades and quotes made at the same points in time.

2. Before reading the data, we will define a schema for it because this information is already known to us. We will assign the PySpark TimestampType class to the columns that hold temporal data about the trade execution and quote time. We will also rename the trade and execution time column names to events_ts to finally convert this data into Delta format. The code to do this is shown in the following snippet:

```
from pyspark.sql.types import *
trade_schema = StructType([
    StructField("symbol", StringType()),
    StructField("event_ts", TimestampType()),
    StructField("trade_dt", StringType()),
    StructField("trade_pr", DoubleType())
])
quote_schema = StructType([
    StructField("symbol", StringType()),
    StructField("event_ts", TimestampType()),
    StructField("trade_dt", StringType()),
    StructField("bid_pr", DoubleType()),
    StructField("ask_pr", DoubleType())
])
```

Once we have downloaded the data and specified the desired schema for our Spark data frame, we can parse the CSV files according to the data schema and store them as Delta tables that we will be able to query afterward.

3. We have decided to use Delta tables to benefit from the optimized data format that allows us to work with large compressed flat files and harness the power of the underlying engine, enabling us to easily scale and parallelize the process according to the amount of data available. The code is illustrated in the following snippet:

```
spark.read.format("csv").schema(trade_schema).
option("header", "true").option("delimiter", ",").load("/
tmp/finserv/ASOF_Trades.csv").write.mode('overwrite').
```

```
format("delta").save('/tmp/finserv/delta/trades')
spark.read.format("csv").schema(quote_schema).
option("header", "true").option("delimiter", ",").load("/
tmp/finserv/ASOF_Quotes.csv").write.mode('overwrite').
format("delta").save('/tmp/finserv/delta/quotes')
```

4. After we have read and stored the data into the Delta tables, we can check that the data has been parsed correctly by displaying the resulting Delta table as a Spark Data Frame, as follows:

```
display(spark.read.format("delta").load("/tmp/finserv/
delta/trades"))
```

Now that we have our transactional data available, we can use it to merge the trades and quotes, aggregate the trades, and perform windowing operations on it.

Joining time-series data

When working with time-series data, an as-of join is a merge technique that commonly refers to obtaining the value of a given event in the exact moment of the timestamp. For most of this temporal data, different types of datetime series will be joined together. In our particular case, we want to know for each company in the dataset the particular state of trade at any given point in time that we have available. Here, these states can be—for example—the **NBBO**, which is an acronym for **National Best Bid Offer** in trading and investing.

In the following example, we will obtain the state of the NBBO for each company available in the dataset. We will work the data in order to have for each company the state of the latest bid and offer for each of the data points that we have as timestamps. Once we have this computed, we can visualize the difference between bids and offers to comprehend at which points in time the liquidity of the company may hit a low point:

1. First, we will merge the data on the symbol. We assume that we have already loaded the Delta tables into two Spark data frames named **trades** and **offers**, which we will merge on the **symbol** column. The code is shown in the following snippet:

```
un= trades.filter(col("symbol") == "K").select('event_
ts', 'price', 'symbol', 'bid', 'offer', 'ind_cd').
union(quotes.filter(col("symbol") == "K").select('event_
ts', 'price', 'symbol', 'bid', 'offer', 'ind_cd'))
```

2. After we have performed the merge, we can visualize the results, as follows:

    ```
    display(un)
    ```

3. Now that our data has been merged into a single Spark DataFrame, we can perform the windowing operation. First, we need to define the windowing function that we will use. To do this, we will use the `Window` class, which is a built-in method in PySpark to perform windowing operations. In the following code snippet, we will define the partition by the symbol column, which represents the company:

    ```
    from pyspark.sql.window import Window
    partition_spec = Window.partitionBy('symbol')
    ```

4. Then, we need to specify the mechanism that will be used to sort the data. Here, we use the `ind_cd` column as the `sort` key, which is the column that specifies the values of the quotes before trades:

    ```
    join_spec = partition_spec.orderBy('event_ts').
    rowsBetween(Window.unboundedPreceding,
                Window.currentRow)
    ```

5. Then, we will use SQL commands to query the data and get the `lasted_bid` value by running a `SELECT` operation over the last bid and over `join_spec`, as follows:

    ```
    %sql
    select(last("bid", True).over(join_spec).alias("latest_
    bid"))
    ```

This way, we have demonstrated how easy is to manipulate time-series data in Azure Databricks using the PySpark API to run joins, merges, and aggregations on temporal data. In the next section, we will learn how we can leverage the benefit of the Pandas-like syntax to ease the transition from notebooks to production easier.

Using the Koalas API

When working with time series, it is fairly common to perform tasks related to imputation and removing duplicates. These duplicated records tend to happen when multiple records are inserted with high frequency on the dataset. We can use the Koalas API to perform these operations using a very familiar Pandas-like syntax, as was explored in previous chapters:

1. In the following code example, we will read our Delta table into a Koalas DataFrame and perform deduplication by grouping by the `event_ts` column and then retrieve the maximum value for that point in time:

    ```
    import databricks.koalas as ks
    kdf_src = ks.read_delta("/tmp/finserv/delta/ofi_quotes2")
    grouped_kdf = kdf_src.groupby(['event_ts'],
                                  as_index=False).max()
    grouped_kdf.sort_values(by=['event_ts'])
    grouped_kdf.head()
    ```

2. Now that we have removed the duplicates from our DataFrame, we can perform a shift in the data to create a lagged Koalas Data Frame that will be really helpful to do calculations such as moving average or other statistical trend calculations. As shown in the following code snippet, Koalas makes it easy to get lag or lead values within a window by using the shift method of the Koalas Data Frame:

    ```
    grouped_kdf.set_index('event_ts', inplace=True,
                          drop=True)
    lag_grouped_kdf = grouped_kdf.shift(periods=1,
                                        fill_value=0)
    lag_grouped_kdf.head()
    ```

3. Now that we have lag values computed, we are to be able to merge this dataset with our original DataFrame to have all the data consolidated into a single structure to ease the calculation of statistical trends. In the following code example, we demonstrate how simple it is to merge two different data frames into a single one:

    ```
    lagged = grouped_kdf.merge(lag_grouped_kdf,
                               left_index=True,
                               right_index=True,
                               suffixes=['', '_lag'])
    lagged.head()
    ```

The Koalas API is very helpful for modeling our data and makes it easy to perform the necessary transformations, even when working with the time-series data. In the next section, we will learn how we can deal with possible missing values in our data frames when working in Azure Databricks.

Handling missing values

Real-life data is far from perfect, and cases of having missing values are really common. The mechanisms in which the data has become unavailable are really important to come up with a good imputation strategy. We call imputation the process in which we deal with values that are missing in our data, which in most contexts are represented as NaN values. One of the most important aspects of this is to know which values are missing:

1. In the following code example, we will show how we can find out which columns have missing or null values by summing up all the Boolean output of the Spark `isNull` method by casting this Boolean output to integers:

    ```
    from pyspark.sql.functions import col, sum
    df.select(*(sum(col(c).isNull().cast("int")).alias(c) for
    c in df.columns)).show()
    ```

2. Another alternative would be to use the output of the Spark data frame describe method to filter out the count of missing values in each column and, finally, subtracting the count of rows to get the actual number of missing values, as follows:

    ```
    from pyspark.sql.functions import lit
    rows = df.count()
    summary = df.describe().filter(col("summary") == "count")
    summary.select(*((lit(rows)-col(c)).alias(c) for c in
    df.columns)).show()
    ```

3. Once we have identified the null values, we can use remove all rows that contain null values by using the `na.drop()` method of Spark data frames, as follows:

    ```
    null_df.na.drop().show()
    ```

4. We can also specify that the rows that have a number of null values greater than 2 be dropped, as follows:

    ```
    null_df.na.drop(thresh=2).show()
    ```

5. One of the other available options is to use the how argument of this method. For example, we can specify this parameter as any, which indicates to drop rows having any number of null values, as illustrated in the following code snippet:

```
null_df.na.drop(how='any').show()
```

6. We can also drop null values on a single column using a subset parameter, which accepts list of column names, as illustrated in the following code snippet:

```
null_df.na.drop(subset=['Sales']).show()
```

7. For example, we can specify to drop records that have null values in all of the specified subset of columns, as follows:

```
null_df.na.drop(how='all',subset=['Name','Sales']).show()
```

As mentioned before, one of the available options when dealing with missing data is to perform an imputation of them, which can be understood as filling those null values according to a predefined strategy that we know won't change the actual distribution of the data. It is always better to impute the null values rather than drop data.

8. For example, we can choose to fill the null values with a string that helps us to identify them, as follows:

```
null_df.na.fill('NA').show()
```

9. Or, we can also fill the missing values with an integer value, like this:

```
null_df.na.fill(0).show()
```

One more common option is filling the missing values with the mean or the median calculated from the present values. This way, we try to preserve the actual distribution of the data and try to not seriously affect other calculations such as the column mean. Nevertheless, it is necessary to keep in mind that if we do a new calculation of the mean, we will get another value.

10. Here, we can fill a numeric column with the mean of the average for that particular column, as follows:

```
from pyspark.sql.functions import mean
mean_val=null_df.select(mean(null_df.Sales)).collect()
print(type(mean_val)) #mean_val is a list row object
```

```
print('mean value of Sales', mean_val[0][0])
mean_sales=mean_val[0][0]
```

11. Now that we have calculated the mean value and stored it as a variable named men_
sales, we can use this value to fill the null values in the sales column, as follows:

```
null_df.na.fill(mean_sales,subset=['Sales']).show()
```

In this way, we can handle all possible null or missing values that we find in the data, although the strategy that will be used to handle them depends greatly on the mechanism that created the missing values and the domain knowledge of the problem.

Extracting features from text

Extracting information from text relies on being able to capture the underlying language structure. This means that we intend to capture the meaning and relationship among tokens and the meaning they try to convey within a sentence. These sorts of manipulations and tasks associated with understanding the meaning in text yield a whole branch of an interdisciplinary field called **natural language processing** (**NLP**). Here, we will focus on some examples related to transforming text into numerical features that can be used later on the machine learning and deep learning algorithms using the PySpark API in Azure Databricks.

TF-IDF

Term Frequency-Inverse Document Frequency (**TF-IDF**) is a very commonly used text preprocessing operation to convert sentences into features created based on the relative frequency of the tokens that compose them. The term frequency-inverse is used to create a set of numerical features that are constructed based on how relevant a word is to the context, not only to a document in a collection or corpus.

In Azure Databricks ML, TF-IDF is an operation that is done in two separate parts, the first being the TF (+hashing), followed by the IDF:

- **TF**: HashingTF is a PySpark Transformer class that takes an array of terms and converts them into fixed-length feature vectors. In feature engineering, when we refer to a "set of terms", this might imply that we are referencing a **bag of words** (**BOW**). The algorithm combines TF counts with the hashing trick (sometimes called **feature hashing**), in order to do a fast and efficient vectorization of the features into vectors or matrices, for dimensionality reduction.

- **IDF**: IDF is a PySpark Estimator class that is fitted on a dataset and yields as a result an IDFModel. The IDFModel takes as input feature vectors, which in this case come from the HashingTF class, and scales each feature column. It will automatically lower weights applied on columns, depending on the frequency it appears on the corpus.

In the following example, we take as input an array of sentences. Each sentence will be split into tokens using the PySpark Tokenizer class, which will yield a BOW for each sentence:

1. Later—and, as mentioned, in the following code block —we will apply the HashingTF class to reduce the dimensionality of the feature vector:

```
from pyspark.ml.feature import HashingTF, IDF, Tokenizer
sentenceData = sqlContext.createDataFrame([
    (0, "Hi I heard about Spark"),
    (0, "I wish Java could use case classes"),
    (1, "Logistic regression models are neat")
], ["label", "sentence"])
tokenizer = Tokenizer(inputCol="sentence",
outputCol="words")
wordsData = tokenizer.transform(sentenceData)
hashingTF = HashingTF(inputCol="words",
                      outputCol="rawFeatures",
                      numFeatures=20)
```

2. Later, we will apply the IDF class to rescale the obtained features as a good practice when we use this in a machine learning or deep learning algorithm to avoid issues with gradients and generally improve the performance. The code for this is shown in the following snippet:

```
featurizedData = hashingTF.transform(wordsData)
idf = IDF(inputCol="rawFeatures",outputCol="features")
idfModel = idf.fit(featurizedData)
rescaledData = idfModel.transform(featurizedData)
```

3. Finally, we can display the obtained features, as follows:

```
for features_label in rescaledData.
select("features","label").take(3):
    print(features_label)
```

4. We can see the entire process of generating the variables, as shown in the following screenshot:

```
Cmd 12

1   from pyspark.ml.feature import HashingTF, IDF, Tokenizer
2   sentenceData = sqlContext.createDataFrame([
3     (0, "Hi I heard about Spark"),
4     (0, "I wish Java could use case classes"),
5     (1, "Logistic regression models are neat")
6   ], ["label", "sentence"])
7   tokenizer = Tokenizer(inputCol="sentence", outputCol="words")
8   wordsData = tokenizer.transform(sentenceData)
9   hashingTF = HashingTF(inputCol="words", outputCol="rawFeatures", numFeatures=20)
10  featurizedData = hashingTF.transform(wordsData)
11  idf = IDF(inputCol="rawFeatures", outputCol="features")
12  idfModel = idf.fit(featurizedData)
13  rescaledData = idfModel.transform(featurizedData)
14  for features_label in rescaledData.select("features","label").take(3):
15    print(features_label)

▸ (3) Spark Jobs
▸ 🖽 sentenceData: pyspark.sql.dataframe.DataFrame = [label: long, sentence: string]
▸ 🖽 wordsData: pyspark.sql.dataframe.DataFrame = [label: long, sentence: string ... 1 more fields]
▸ 🖽 featurizedData: pyspark.sql.dataframe.DataFrame = [label: long, sentence: string ... 2 more fields]
▸ 🖽 rescaledData: pyspark.sql.dataframe.DataFrame = [label: long, sentence: string ... 3 more fields]
Row(features=SparseVector(20, {6: 0.2877, 8: 0.6931, 13: 0.2877, 16: 0.5754}), label=0)
Row(features=SparseVector(20, {0: 0.6931, 2: 0.6931, 7: 1.3863, 13: 0.2877, 15: 0.6931, 16: 0.2877}), label=0)
Row(features=SparseVector(20, {3: 0.6931, 4: 0.6931, 6: 0.2877, 11: 0.6931, 19: 0.6931}), label=1)
```

Figure 8.8 – Output features of the TF-IDF operation

In this way, we have our features extracted from the text and converted into scaled numerical features that can be fed into any learning algorithm.

Word2vec

Word2vec is an embedding technique that is used in feature engineering for NLP. It uses a neural network to learn the associations between words from a large corpus of text and it produces vectors that represent each token. This vector holds properties that allow us to indicate the semantic similarity between two words just by measuring the distance between the vectors that represent them, using (for example) cosine similarity. These vectors are often called word embeddings and they commonly have several hundreds of dimensions used to represent them.

In PySpark, `Word2vec` is an `Estimator` class that takes as input an array of and trains a `Word2VecModel`, which is a model that maps each word into a single fixed-sized vector.

In the code example shown next, we start with a set of documents, each of them represented as a sequence of words. For each document, we transform it into a feature vector. This feature vector could then be passed to a learning algorithm:

1. In the following code example, we will create a Spark data frame with sentences that are split into their tokens using a simple split on the string, meaning that each row represents a BOW from that sentence:

```
from pyspark.ml.feature import Word2Vec
documentDF = sqlContext.createDataFrame([
    ("Hi I heard about Spark".split(" "), ),
    ("I wish Java could use case classes".split(" "), ),
    ("Logistic regression models are neat".split(" "), )
], ["text"])
```

2. Then, we can map those arrays into vectors, as follows:

```
word2Vec = Word2Vec(vectorSize=3, minCount=0,
inputCol="text", outputCol="result")
```

3. Finally, we can fit the Word2Vec model to the array we just created and obtain the resulting features by transforming the input DataFrame using the fitted model, as follows:

```
model = word2Vec.fit(documentDF)
result = model.transform(documentDF)
for feature in result.select("result").take(3):
    print(feature)
```

4. We can see here the whole process executed on a single cell in an Azure Databricks notebook:

Cmd 14

```
1   from pyspark.ml.feature import Word2Vec
2   documentDF = sqlContext.createDataFrame([
3     ("Hi I heard about Spark".split(" "), ),
4     ("I wish Java could use case classes".split(" "), ),
5     ("Logistic regression models are neat".split(" "), )
6   ], ["text"])
7   word2Vec = Word2Vec(vectorSize=3, minCount=0, inputCol="text", outputCol="result")
8   model = word2Vec.fit(documentDF)
9   result = model.transform(documentDF)
10  for feature in result.select("result").take(3):
11    print(feature)
12
```

▸ (4) Spark Jobs

  ▸ ▦ documentDF: pyspark.sql.dataframe.DataFrame = [text: array]

  ▸ ▦ result: pyspark.sql.dataframe.DataFrame = [text: array, result: udt]

```
Row(result=DenseVector([-0.0627, -0.0219, -0.0816]))
Row(result=DenseVector([0.0242, 0.0236, 0.023]))
Row(result=DenseVector([0.0483, -0.0189, -0.0037]))
```

Figure 8.9 – Output embeddings using Word2Vec

In this way, we can very simply create a numerical representation obtained from text sources, using one of the most widely used models in text embeddings: Word2Vec.

Training machine learning models on tabular data

In this example, we will use a very popular dataset in data science, which is the wine dataset of physicochemical properties, to predict the quality of a specific wine. We will be using Azure Databricks Runtime ML, so be sure to attach the notebook to a cluster running this version of the available runtimes, as specified in the requirements at the beginning of the chapter.

Engineering the variables

We'll get started using the following steps:

1. Our first step is to load the necessary data to train our models. We will load the datasets, which are stored as example datasets in DBFS, but you can also get them from the UCI Machine Learning repository. The code is shown in the following snippet:

    ```
    import pandas as pd
    white_wine = pd.read_csv("/dbfs/databricks-datasets/wine-
    quality/winequality-white.csv", sep=";")
    red_wine = pd.read_csv("/dbfs/databricks-datasets/wine-
    quality/winequality-red.csv", sep=";")
    ```

2. Next, we will merge these two pandas Data Frames into a single one, and add a new binary feature named is_red to distinguish between the two of them, as follows:

    ```
    red_wine['is_red'] = 1
    white_wine['is_red'] = 0
    data = pd.concat([red_wine, white_wine], axis=0)
    ```

3. After this, we will remove spaces from the column names in order to be able to save this as a Spark data frame, as follows:

    ```
    data.rename(columns=lambda x: x.replace(' ', '_'),
    inplace=True)
    data.head()
    ```

4. Here, we can see the created Pandas Data frame:

Cmd 17

```
1  data.rename(columns=lambda x: x.replace(' ', '_'), inplace=True)
2  data.head()
```

Out[30]:

	fixed_acidity	volatile_acidity	citric_acid	residual_sugar	chlorides	free_sulfur_dioxide	total_sulfur_dioxide	density	pH	sulphates	alcohol	quality	is_red
0	7.4	0.70	0.00	1.9	0.076	11.0	34.0	0.9978	3.51	0.56	9.4	5	1
1	7.8	0.88	0.00	2.6	0.098	25.0	67.0	0.9968	3.20	0.68	9.8	5	1
2	7.8	0.76	0.04	2.3	0.092	15.0	54.0	0.9970	3.26	0.65	9.8	5	1
3	11.2	0.28	0.56	1.9	0.075	17.0	60.0	0.9980	3.16	0.58	9.8	6	1
4	7.4	0.70	0.00	1.9	0.076	11.0	34.0	0.9978	3.51	0.56	9.4	5	1

Figure 8.10 – Renamed columns of the dataset

5. Now that we have a unified dataset, we will create a new feature named `high_quality` for all the wines that have a quality above 7, as follows:

```
high_quality = (data.quality >= 7).astype(int)
data.quality = high_quality
```

6. Afterward, we can split our dataset in train and test datasets, which we will use to train our machine learning model later on, inferring on the quality column. In the following code block, we import the scikit-learn `train_test_split` function in order to split the data into the features defined by X and the objective column defined as y:

```
from sklearn.model_selection import train_test_split
train, test = train_test_split(data, random_state=123)
X_train = train.drop(["quality"], axis=1)
X_test = test.drop(["quality"], axis=1)
y_train = train.quality
y_test = test.quality
```

As a result, we have the features and objective columns split into train and test datasets that can be used to train and benchmark the model.

Building the ML model

Here, we will use a classifier given the fact that what we are trying to predict is a binary output on a tabular dataset. The selected algorithm will be a Random Forest classifier available in the `scikit-learn` Python library. We will also make use of `MLflow` to keep track of the model performance, and to save the model for later use:

1. We will start by doing all the necessary imports, as follows:

```
import mlflow
import mlflow.pyfunc
import mlflow.sklearn
import numpy as np
from sklearn.ensemble import RandomForestClassifier
from sklearn.metrics import roc_auc_score
from mlflow.models.signature import infer_signature
```

2. We will create a wrapper class named `SklearnModelWrapper` around the model, inheriting from the `mlflow.pyfunc.PythonModel` class. This class will return the probability of an instance being part of a determined class. The code is shown in the following snippet:

```
class SklearnModelWrapper(mlflow.pyfunc.PythonModel):
  def __init__(self, model):
    self.model = model
  def predict(self, context, model_input):
    return self.model.predict_proba(model_input)[:,1]
```

3. After we have defined our model, we will use `mlflow.start_run` to create a new MLflow run that will allow us to keep up with the track performance. We can also call `mlflow.log_param` to show which parameters are being passed to the model. In the following code example, `mlflow.log_metric` will be recording the performance metrics that we have selected, such as the **receiver operating characteristic (ROC)** in this case:

```
with mlflow.start_run(run_name='untuned_random_forest'):
  n_estimators = 10
  model = RandomForestClassifier(n_estimators=n_
estimators, random_state=np.random.RandomState(123))
  model.fit(X_train, y_train)
  predictions_test = model.predict_proba(X_test)[:,1]
  auc_score = roc_auc_score(y_test, predictions_test)
  mlflow.log_param('n_estimators', n_estimators)
  mlflow.log_metric('auc', auc_score)
  wrappedModel = SklearnModelWrapper(model)
  signature = infer_signature(X_train, wrappedModel.
predict(None, X_train))
  mlflow.pyfunc.log_model("random_forest_model", python_
model=wrappedModel, signature=signature)
```

4. In the preceding code snippet, the `predict_proba` function will return a tuple with the probability of belonging to one of the two classes. We will keep just one, so we slice the output with `[:, 1`. After the model has been deployed, we create a signature that will be used later on to validate inputs. Here, we can see the whole code being run on an Azure Databricks notebook:

Cmd 21

```python
1   import mlflow
2   import mlflow.pyfunc
3   import mlflow.sklearn
4   import numpy as np
5   from sklearn.ensemble import RandomForestClassifier
6   from sklearn.metrics import roc_auc_score
7   from mlflow.models.signature import infer_signature
8
9   class SklearnModelWrapper(mlflow.pyfunc.PythonModel):
10    def __init__(self, model):
11      self.model = model
12
13    def predict(self, context, model_input):
14      return self.model.predict_proba(model_input)[:,1]
15
16  with mlflow.start_run(run_name='untuned_random_forest'):
17    n_estimators = 10
18    model = RandomForestClassifier(n_estimators=n_estimators, random_state=np.random.RandomState(123))
19    model.fit(X_train, y_train)
20    predictions_test = model.predict_proba(X_test)[:,1]
21    auc_score = roc_auc_score(y_test, predictions_test)
22    mlflow.log_param('n_estimators', n_estimators)
23    mlflow.log_metric('auc', auc_score)
24    wrappedModel = SklearnModelWrapper(model)
25    signature = infer_signature(X_train, wrappedModel.predict(None, X_train))
26    mlflow.pyfunc.log_model("random_forest_model", python_model=wrappedModel, signature=signature)
27
```

```
/databricks/python/lib/python3.7/site-packages/mlflow/models/signature.py:123: UserWarning: Hint: Inferred schem
represent missing values. If your input data contains missing values at inference time, it will be encoded as fl
avoid this problem is to infer the model schema based on a realistic data sample (training dataset) that include
s as doubles (float64) whenever these columns may have missing values. See `Handling Integers With Missing Value
gers-with-missing-values>`_ for more details.
  inputs = _infer_schema(model_input)
Command took 1.82 seconds --
```

Figure 8.11 – Fitting the model using the MLflow start_run decorator

5. After the model has been trained, we can examine the learned feature importance output of the model as a sanity check, as follows:

```
feature_importances = pd.DataFrame(model.feature_
importances_, index=X_train.columns.tolist(),
columns=['importance'])
```

```
feature_importances.sort_values('importance',
ascending=False)
```

6. Here, we can see which variables were most important for the model:

Figure 8.12 – Model feature importance

7. It shows that both alcohol and density are important in predicting quality. You can click on the **Experiment** button at the upper right to see the **Experiment Runs** sidebar, as illustrated in the following screenshot:

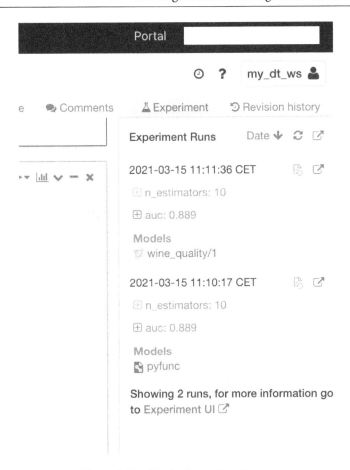

Figure 8.13 – Tracked experiment runs

In this example, this model achieved an **area under the curve** (**AUC**) of **0.889**.

Registering the model in the MLflow Model Registry

The next steps will be to register the model that we just trained in the MLflow Model Registry, which is a centralized model registry that allows us to manage the full life cycle of a trained model and to use our model from anywhere in Azure Databricks. This model repository will not only store the models but will also provide us with versioning control of them, as well as allow us to add descriptions and comments:

1. In the next example, we will show how we can register our model programmatically, but you can also do this by using the **user interface (UI)** in the Azure Databricks workspace. First, we will store the run ID of the latest run for the model that we just trained. We search for the run ID, filtering by the name of the model using the `tags.mlflow.runName` parameter, as follows:

    ```
    run_id = mlflow.search_runs(filter_string='tags.mlflow.
    runName = "untuned_random_forest"').iloc[0].run_id
    model_name = "wine_quality"
    model_version = mlflow.register_model(f"runs:/{run_id}/
    random_forest_model", model_name)
    ```

    If you navigate to the **Models** page in your Azure Databricks workspace, you will see the model we just registered.

2. After our model has been successfully registered, we can set the stage of this model. We will move our model to production and load it in the notebook, as shown in the following code snippet:

    ```
    from mlflow.tracking import MlflowClient
    client = MlflowClient()
    client.transition_model_version_stage(
     name=model_name,
     version=model_version.version,
     stage="Production",
    )
    ```

3. We can see here in the cell output that the model was correctly registered:

```
Cmd 23

1  run_id = mlflow.search_runs(filter_string='tags.mlflow.runName = "untuned_random_forest"').iloc[0].run_id
2  model_name = "wine_quality"
3  model_version = mlflow.register_model(f"runs:/{run_id}/random_forest_model", model_name)


Successfully registered model 'wine_quality'.
2021/03/15 10:14:16 INFO mlflow.tracking._model_registry.client: Waiting up to 300 seconds for model version to finish creation.
version 1
Created version '1' of model 'wine_quality'.
```

Figure 8.14 – Correctly registered model

4. Now, we can use our model by referencing using the `f"models:/{model_name}/production"` path. The model can then be loaded into a Spark **user-defined function (UDF)** so that it can be applied to the Delta table for batch inference. The following code will load our model and apply it to a Delta table:

```
import mlflow.pyfunc
from pyspark.sql.functions import struct
apply_model_udf = mlflow.pyfunc.spark_udf(spark,
f"models:/{model_name}/production")
new_data = spark.read.format("delta").load(table_path)
```

5. Once the model has been loaded and the data has been read from the Delta table, we can use a struct to pass the input variables to the model registered as a UDF, as follows:

```
udf_inputs = struct(*(X_train.columns.tolist()))
new_data = new_data.withColumn(
  "prediction",
  apply_model_udf(udf_inputs)
)
display(new_data)
```

Here, each record in the Delta table will have now an associated prediction.

Model serving

We can use our model for low-latency predictions by using the `MLflow` model serving to provide us with an endpoint so that we can issue requests using a REST API and get the predictions as a response.

You will need your Databricks token to be able to issue the request to the endpoint. As mentioned in previous chapters, you can get your token in the **User Settings** page in Azure Databricks.

Before being able to make requests to our endpoint, remember to enable the serving on the model page in the Azure Databricks workspace. You can see a reminder of how to do this in the following screenshot:

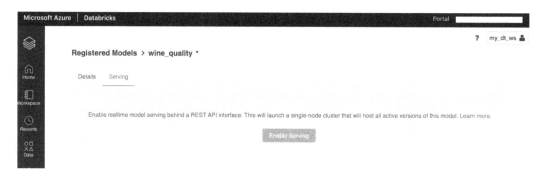

Figure 8.15 – Enabling model serving

1. Finally, we can call our model. The following example function will issue a Pandas data frame and will return the inference for each row in the DataFrame:

```
import os
import requests
import pandas as pd
def score_model(dataset: pd.DataFrame):
  url = 'https://YOUR_DATABRICKS_URL/model/wine_quality/
Production/invocations'
  headers = {'Authorization': f'Bearer {your_databricks_
token}'
  data_json = dataset.to_dict(orient='split')
  response = requests.request(method='POST',
headers=headers, url=url, json=data_json)
  if response.status_code != 200:
    raise Exception(f'Request failed with status
{response.status_code}, {response.text}')
  return response.json()
```

2. We can compare the results we get on the local model by passing the X_test data, as shown in the following code snippet. We should obtain the same results:

```
num_predictions = 5
served_predictions = score_model(X_test[:num_
predictions])
model_evaluations = model.predict(X_test[:num_
predictions])
pd.DataFrame({
  "Model Prediction": model_evaluations,
  "Served Model Prediction": served_predictions,
})
```

This way, we have enabled low-latency predictions on smaller batches of data by using our published model as an endpoint to issue requests.

Summary

In this section, we have covered many examples related to how we can extract and improve features that we have available in the data, using methods such as tokenization, polynomial expansion, and one-hot encoding, among others. These methods allow us to prepare our variables for the training of our models and are considered as a part of feature engineering.

Next, we dived into how we can extract features from text using TF-IDF and Word2Vec and how we can handle missing data in Azure Databricks using the PySpark API. Finally, we have finished with an example of how we can train a deep learning model and have it ready for serving and get predictions when posting REST API requests.

In the next chapter, we will focus more on handling large amounts of data for deep learning using TFRecords and Petastorm, as well as on how we can leverage existing models to extract features from new data in Azure Databricks.

9
Databricks Runtime for Deep Learning

This chapter will take a deep dive into the development of classic deep learning algorithms to train and deploy models based on unstructured data, exploring libraries and algorithms as well. The examples will be focused on the particularities and advantages of using Databricks for DL, creating DL models. In this chapter, we will learn about how we can efficiently train deep learning models in Azure Databricks and implementations of the different libraries that we have available to use.

The following topics will be introduced in this chapter:

- Loading data for deep learning
- Managing data using TFRecords
- Automating scheme inference
- Using Petastorm for distributed learning
- Reading a dataset
- Data preprocessing and featurization

This chapter will have more of a focus on deep learning models rather than machine learning ones. The main distinction is that we will focus more on handling large amounts of unstructured data. This chapter will end with an example of how we can use a pre-trained model in TensorFlow to extract features based on new domain data.

Technical requirements

To work on the examples given in this chapter, it is required for you to have the following:

- An Azure Databricks subscription
- An Azure Databricks notebook attached to a running cluster with Databricks Runtime ML of version 7.0 or higher.

Loading data for deep learning

In this chapter, we will learn how we can prepare data for distributed training. To do this, we will learn how to efficiently load data to create deep learning based applications that can leverage the distributed computing nature of Azure Databricks while handling large amounts of data. We will describe two different methods that we have at our disposal for working with large datasets for distributed training. Those methods are Petastorm and TFRecord, which are libraries that make our work easier when loading large and complex datasets to our deep learning algorithms in Azure Databricks.

At a quick glance, the main characteristics of the Petastorm and TFRecord methods are as follows:

- **Petastorm**: It is an open source library that allows us to directly load data in Apache Parquet format to train our deep learning algorithms. This is a great feature of Azure Databricks because Parquet is a widely used format when working with large amounts of data.

- **TFRecord**: Another alternative is to load our data using the TFRecord format as the data source for distributed deep learning in Azure Databricks, which is based on the TensorFlow framework. The TFRecord format is a simple record-oriented binary format that can be used to store data for TensorFlow applications or to store training data. In the TensorFlow library, `tf.data.TFRecordDataset` is a class used for that, which in turn is composed of records from TFRecord files.

In the following sections of this chapter, we will describe and illustrate the recommended ways to save your data to TFRecord files and load them back to train our deep learning TensorFlow models, as well as how we can use Petastorm to read Parquet files for training deep learning models in Azure Databricks.

Using TFRecords for distributed learning

Although it is still not such a commonly used component of TensorFlow, the TFRecord file format is TensorFlow's own binary storage format used to store data for training our models. The binary format is an advantage when working with large datasets because it can significantly improve the performance of our processing and loading pipeline, which in turn reduces the required time to train our models. This binary data is more efficiently stored, occupies less space, and takes less time to read and write.

The TFRecord format also allows us to take advantage of multiprocessing and distributed computing features of systems like Azure Databricks, which is commonly used when dealing with large amounts of data that are too big to be stored fully in memory. In these cases, when the amount of data is too big to be stored in memory, the binary format becomes especially useful as it reduces the memory footprint of the data and the time taken to process it, and then this process is done in batches.

Another one of the features of TFRecord is that it is also optimized for storing sequential data such as time series or word embeddings, which is very useful in the case of sales trends or natural language processing applications.

In the next sections, we will learn how we can structure TFRecords, how we use them to handle large amounts of data, and how we can optimize their loading process by specifying the schema inference, all of which allows us to efficiently train our models in Azure Databricks.

Structuring TFRecords files

The TFRecord file format is a simple record-oriented binary format that was designed with the goal of more efficiently storing data for machine and deep learning model training. The TFRecordDataset class allows us to stream files in the TFRecord format as part of our training pipeline. These TFRecord files store the data as a sequence of binary strings, which makes it necessary to predefine the structure of the data before it can be read or written from a file. To specify the structure of the data, we can use either the tf.train.Example or the tf.train.SequenceExample TensorFlow class to store samples of data, after which you must serialize it, and then it can be written to disk using TFRecordWriter.

One of the things that makes this library more optimized is that tf.train.Example is not a common Python class but a protocol buffer, which is a method to serialize structured data more efficiently.

In Azure Databricks, we can use the `spark-tensorflow-connector` library to leverage the advantages of the TFRecord file format and use it to save Apache Spark DataFrames to TFRecord files. This library allows for the conversion of Spark data frames to TFRecords by simply using PySpark APIs and is installed by default in the Azure Databricks Machine Learning Runtime, which is a runtime that is optimized for machine and deep learning applications. If you are not using this version of the runtime, you will need to manually install the library from the Maven repository to the cluster or notebook as we have seen previously in other chapters.

The `spark-tensorflow-connector` library enables us to read TFRecord files in distributed file systems such as Azure Databricks Spark data frames through the PySpark API. When we read our data, we can pass the following options to the parser:

- **load**: It specifies the input path for the TFRecord file and, similar to Spark, it can accept standard globbing expressions.

- **schema**: It is the schema of the TFRecord file. Optionally, we can define this schema using Spark `StructType`. If nothing is passed, the schema will be inferred from the TFRecord file.

- **recordType**: This is the input format of the TFRecord file and accepts two possible options, which are `Example` and `SequenceExample`. The default option is `Example`.

When we write a Spark Dataframe as a TFRecord file, we can pass the following options to the API:

- **save**: This is the output path to where we will write the TFRecord file and can be either a local or distributed filesystem location.

- **codec**: This is the codec that will be used to compress the TFRecord file. For example, if we specify the option `org.apache.hadoop.io.compress.GzipCodec`, it enables gzip compression. While reading compressed TFRecord files, the codec that was used to compress them can be inferred automatically, so this is not a required option.

- **recordType**: This is the desired output format of the TFRecord file. Likewise, as when we read, the available options are `Example` and `SequenceExample` and the default option is `Example`.

- **writeLocality**: It will determine whether the TFRecord file will be written locally on the workers or on a distributed filesystem. The possible options are as follows:

  a) **distributed**: This is the default option and determines that the file is written using Spark's default filesystem.

  b) **local**: This option writes the file on the disks of each of the workers, in a partitioned manner, writing a subset of data on each worker's local disk.

When we use the local mode, it will write the data onto each of the worker's stores local disks a subset of the data. Which subset of data will be stored on which worker will depend on how the data frame was partitioned. Later, these subsets of data will be merged into a single TFRecord file and written on the node where the partition lives. This is a feature that becomes useful when we train our machine and deep learning algorithms in a distributed way, where each worker receives a subset of data to train on. When the local mode is specified as an option, the path given to the writer will be understood as the base path on each of the workers where the subsets of data will be written.

Managing data using TFRecords

In this section, we will demonstrate how to save image data from Spark DataFrames to TFRecords and load it using TensorFlow in Azure Databricks. We will use as an example the flowers example dataset available in the Databricks filesystem, which contains flower photos stored under five sub-directories, one per class.

We will load the flowers Delta table, which contains the preprocessed flowers dataset using a binary file data source and stored as a Spark DataFrame. We will use this data to demonstrate how you can save data from Spark DataFrames to TFRecords:

1. As the first step, we will load the data using PySpark:

```
from pyspark.sql.functions import col
import tensorflow as tf
spark_df = spark.read.format("delta").load("/databricks-
datasets/flowers/delta") \
    .select(col("content"), col("label_index")) \
    .limit(100)
```

2. Next, we will save the loaded data to TFRecord-type files:

```
path = '/ml/flowersData/converted_data.tfrecord'
spark_df.write.format("tfrecords").mode("overwrite").
save(path)
display(dbutils.fs.ls(path))
```

As you can see, we can easily save Spark DataFrames to TFRecord files, thanks to the underlying spark-tensorflow-connector library, which integrates Spark with TensorFlow:

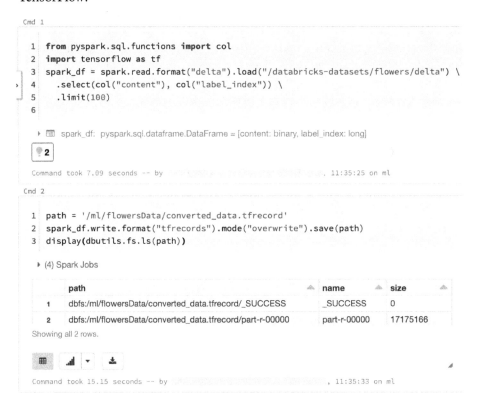

Fig.9.1 – Output location of the tfrecord file

3. The next step in learning how to use TFRecords in Azure Databricks is to load the TFRecord data into TensorFlow. First, we will create a `TFRecordDataset` as an input pipeline:

```
import os
filenames = [(f"/dbfs{path}/{name}") for name in
os.listdir("/dbfs" + path) if name.startswith("part")]
```

```
dataset = tf.data.TFRecordDataset(filenames)
```

4. To be able to use the data, we need to create a decoder function to read, parse, and normalize the data. We parse the record into tensors with the map function, which will apply our decoder function to every sample. First, we will define the decoder function, which will take one serialized example as a required parameter, and the image size as an optional parameter:

```python
def decode_and_normalize(serialized_example, image_size =
224):
  # Parse from single example
  feature_dataset = tf.io.parse_single_example(
      serialized_example,
      features={
          'content': tf.io.FixedLenFeature([],
tf.string),
          'label_index': tf.io.FixedLenFeature([],
tf.int64),
      })
  # Decode the parsed data
  image = \
  tf.io.decode_jpeg(feature_dataset['content'])
  label = \
  tf.cast(feature_dataset['label_index'], tf.int32)
  # Resize the decoded data into the desired size
  image = tf.image.resize(image, [image_size,
                                  image_size])
  # Finally, normalize the data
  image = \
  tf.cast(image, tf.float32) * (1. / 255) - 0.5
  return image, label
```

5. Once we have defined our decoding function, we can use it to parse our data by mapping the dataset and applying it to every sample in it:

```python
parsed_dataset = dataset.map(decode_and_normalize)
```

This way, we have parsed the dataset and now it can be used as input data to train machine or deep learning models.

In the next sections, we dive into a full example of how to use TFRecord in Azure Databricks.

Automating schema inference

The spark-tensorflow-connector library, which integrates Spark with TensorFlow, supports automatic schema inference when reading TensorFlow records into Spark DataFrames. Schema inference is an expensive operation because it requires an extra reading pass through the data, and therefore it's good practice to specify it as it will improve the overall performance of our pipeline.

The following Python code example demonstrates how we can do this on some test data we create as an example:

1. Our first step is to define the schema of our data:

    ```python
    from pyspark.sql.types import *
    path = "test-output.tfrecord"
    fields = [StructField("id", IntegerType()),
    StructField("IntegerCol", IntegerType()),
    StructField("LongCol", LongType()),
    StructField("FloatCol", FloatType()),
    StructField("DoubleCol", DoubleType()),
    StructField("VectorCol", ArrayType(DoubleType(),
              True)),
    StructField("StringCol", StringType())]
    schema = StructType(fields)
    ```

2. Next, we create a Spark DataFrame with test data, parallelize it, and write it as a TFRecord file:

    ```python
    test_rows = [[11, 1, 23, 10.0, 14.0, [1.0, 2.0], "r1"],
    [21, 2, 24, 12.0, 15.0, [2.0, 2.0], "r2"]]
    rdd = spark.sparkContext.parallelize(test_rows)
    df = spark.createDataFrame(rdd, schema)
    ```

```
df.write.format("tfrecords").option("recordType",
"Example").save(path)
```

After this, we can read the TFRecord file back to a Spark Data frame in the next way:

```
df = spark.read.format("tfrecords").option("recordType",
"Example").load(path)
df.show()
```

3. Here, we see the entire process running in an Azure Databricks notebook:

```
Cmd 7
1  from pyspark.sql.types import *
2  path = "test-output.tfrecord"
3  fields = [StructField("id", IntegerType()),
4  StructField("IntegerCol", IntegerType()),
5  StructField("LongCol", LongType()),
6  StructField("FloatCol", FloatType()),
7  StructField("DoubleCol", DoubleType()),
8  StructField("VectorCol", ArrayType(DoubleType(), True)),
9  StructField("StringCol", StringType())]
10 schema = StructType(fields)
11
12 test_rows = [[11, 1, 23, 10.0, 14.0, [1.0, 2.0], "r1"], [21, 2, 24, 12.0, 15.0, [2.0, 2.0], "r2"]]
13 rdd = spark.sparkContext.parallelize(test_rows)
14 df = spark.createDataFrame(rdd, schema)
15 path= 'dbfs:/tmp/dataset'
16 df.write.format("tfrecords").option("recordType", "Example").save(path)
17 display(df)
```

▸ (3) Spark Jobs

▸ 🗎 df: pyspark.sql.dataframe.DataFrame = [id: integer, IntegerCol: integer ... 5 more fields]

	id	IntegerCol	LongCol	FloatCol	DoubleCol	VectorCol	StringCol
1	11	1	23	10	14	▸ [1, 2]	r1
2	21	2	24	12	15	▸ [2, 2]	r2

Showing all 2 rows.

Fig.9.2 – TFRecord column example data types

As you can see, the integration of Spark with the TFRecord file format in Azure Databricks is seamless and allows us to read and write TFRecord files directly in Spark DataFrames.

Using TFRecordDataset to load data

Another of the available options that we have for working with TFRecord files is to use the TFRecordDataset TensorFlow class. This can be useful when we need to standardize input data to improve the performance of our loading process:

1. You can also pass a list with the file paths we want the TFRecordDataset class to read, as shown in the next code block:

    ```
    filenames = [<your_tfrecord_file_1>,..., [<your_tfrecord_
    file_n>]
    raw_dataset = tf.data.TFRecordDataset(filenames)
    ```

    raw_dataset is a TFRecordDataset class that contains serialized tf.train.
    Example records. When we iterate over it, it will return these records as scalar string tensors.

2. In the next example, we will use the take method to only show the first 10 records in raw_dataset:

    ```
    for raw_record in raw_dataset.take(10):
      print(repr(raw_record))
    ```

    Remember that to iterate over TFRecordDataset, it is necessary to enable eager execution.

3. The tf.train.Example scalar string tensors can then be parsed using the tf.io.FixedLenFeature function. Notice that we need to define feature_description because datasets use graph execution, and this description is used to build their shape and type signature:

    ```
    feature_description = {
        'feature0': tf.io.FixedLenFeature([], tf.int64,
    default_value=0),
        'feature1': tf.io.FixedLenFeature([], tf.string,
    default_value=''),
        'feature2': tf.io.FixedLenFeature([], tf.float32,
    default_value=0.0),
    }
    ```

`feature_description` is a dictionary that holds the description of the features, which is necessary for our parsing function.

4. Here, our next step will define the parsing function, which will read the `tf.train.Example` class input using the previously defined `feature_description` dictionary:

```
def _parsing_function(proto_input):
    return tf.io.parse_single_example(proto_input,
feature_description)
```

5. As the next step to read our TFRecord files, we will apply the parsing function that we have defined to each item in the dataset using the `tf.data.Dataset.map` method:

```
parsed_dataset = raw_dataset.map(_parsing_function)
parsed_dataset
```

We can also parse batches of data using the `tf.io.parse_example` function as an alternative to reading them one by one.

6. We can examine the data by displaying the observations in the dataset when eager execution is enabled. Here, regardless of the number of observations in the dataset, we will only display the first 10 of them, which will be shown as a dictionary of features:

```
for parsed_record in parsed_dataset.take(10):
    print(repr(parsed_record))
```

Each item shown is a `tf.Tensor`, which is a TensorFlow class that has as a value the numpy element of this tensor.

7. To write observations as a TFRecord file, we need to transform each observation to a `tf.train.Example` class before it is written to file. Here, we will write n_obs number of `tf.train.Example` records to a TFRecord file:

```
with tf.io.TFRecordWriter(filename) as writer:
    for i in range(n_obs):
        example = serialize_example(feature0[i],
                                    feature1[i],
                                    feature2[i])
        writer.write(example)
```

The TFRecord file format optimizes our training pipelines when working with large amounts of data that need to be processed using parallel or distributed methods. As shown in this section, we can use native tools of TensorFlow to manipulate TFRecord files to leverage the advantages of distributed computing systems such as Azure Databricks. In the next section, we will introduce Petastorm, a library to enable the use of Parquet files as training data for machine and deep learning applications.

Using Petastorm for distributed learning

Petastorm is an open source library that allows us to do single or distributed training of machine and deep learning algorithms using datasets stored as Apache Parquet files. It supports popular frameworks such as PyTorch, TensorFlow, and PySpark and can also be used for other Python applications. Petastorm provides us with a simple function to augment the functionality of the Parquet format with Petastorm-specific data to be able to be used in machine and deep learning model training. We can simply read our data by creating a reader object from Databricks File System and iterating over it. The underlying Petastorm library uses the PyArrow library to read Parquet files.

In this section, we will discuss how we can use Petastorm to further extend the performance of our machine and deep learning training pipelines in Azure Databricks.

Introducing Petastorm

As mentioned before, Petastorm is an open source library that enables a single machine or distributed training and evaluation of machine and deep learning models using data stored in Apache Parquet format as well as data already loaded in Spark DataFrames. This library supports popular data science frameworks such as TensorFlow, PyTorch, and PySpark. Petastorm is the best choice when we need to train machine and deep learning models using data stored in Parquet format.

The Petastorm library has various features such as optimized data sharing, data partitioning, shuffling and row filtering, and support for time series using n-grams.

N-grams are used when the temporal context is required for models that make use of observed environment variables to better understand the environment and predict future events in that environment.

Petastorm can provide this temporal context when the input data is a time series. When an n-gram is requested from the Petastorm Reader object, sequential rows will be merged into a single training sample to provide the temporal context while keeping the data structured.

When loading data using Petastorm, we have two available options to parallelize the data loading and decoding operations:

- **Thread pool**: This implementation is used when the observations are encoded data or high-resolution images. When this mode is selected, most of the processing time is spent decoding the images through C++ code with almost no Python code.

- **Process pool**: This implementation is appropriate when row sizes are small. When this mode is selected, most decoding is done using only Python code in more than one process running in parallel.

The options selected for parallelizing the data loading and decoding operations will depend on the kind of data you want to read.

As we have introduced in this section, Petastorm is an open source library that helps us leverage the efficiency of the Parquet file format and allows us to use it to train deep learning models in a distributed manner. In the next sections, we will learn how to use Petastorm to create a dataset and use it to prepare image data to train our deep learning models in the Azure Databricks Machine Learning Runtime.

Generating a dataset

Similarly, as we do for other dataset file formats, when we want to generate a dataset using Petastorm, we need to first define a data schema, which is named Unischema. This is the only time that we need to do this as Petastorm will then broadcast this to all supported frameworks, such as PySpark, TensorFlow, and pure Python.

The defined Unischema is serialized as a custom field in the Parquet store metadata, therefore it can be read by just specifying the path to the dataset.

The following example shows how we can create a Unischema instance that will be used to define the schema of our dataset. To properly define the schema, the required parameters to be passed in the field properties are the field name, the data type represented as a NumPy data type, the shape of the array, the codec that will be used to encode and decode the data, and whether a field is nullable or not:

```
first_schema = Unischema('first_schema', [
   UnischemaField('id', np.int32, (),
ScalarCodec(IntegerType()), False),
   UnischemaField('image1', np.uint8, (128, 256, 3)
CompressedImageCodec('png'), False),
```

```
  UnischemaField('array_4d', np.uint8, (None, 128, 30, None),
NdarrayCodec(), False),
])
```

We use can also use the PySpark API to write the Petastorm dataset. The following code example demonstrates how we can create a 10-row dataset using Petastorm and the previously defined data schema:

```
rows_count = 10
with materialize_dataset(spark, output_url, 'first_schema',
rowgroup_size_mb):
    rows_rdd = sc.parallelize(range(rows_count))\
        .map(row_generator)\
        .map(lambda x: dict_to_spark_row(first_schema', x))
    spark.createDataFrame(rows_rdd,
                            first_schema'.as_spark_schema()) \
        .write \
        .parquet('file:///tmp/example_dataset')
```

Let's explain a bit more about what is going on in this example. `materialize_dataset` is the context manager, and it will take care of all the necessary configurations based on the specified data schema and finally write out the Petastorm-specific metadata. The output file path could point either locally or to a location in Databricks File System. `rowgroup_size_mb` defines the target size of the Parquet row group in megabytes and `row_generator` is a function that returns a Python dictionary matching the specified data schema. The `dict_to_spark_row` will validate the data types according to the specified schema and convert the dictionary into a `pyspark.Row` object.

We can load data from Spark DataFrames using Petastorm using the Petastorm Spark converter API, which enables us to convert Spark DataFrames to TensorFlow or PyTorch formats. This is done by first writing the input Spark DataFrame into a Parquet file that then can be loaded either using the `tf.data.Dataset` or `torch.utils.data.DataLoader`.

When we are distributed training deep learning models, the recommended workflow is to load and process the data in Azure Databricks, using the Petastorm `spark_dataset_converter` method to convert data from a Spark DataFrame to a TensorFlow dataset or a PyTorch DataLoader, and finally using it to train our deep learning model.

Reading a dataset

Reading datasets using Petastorm can be very simple. In this section, we will demonstrate how we can easily load a Petastorm dataset into two frequently used deep learning frameworks, which are TensorFlow and PyTorch:

1. To load our Petastorm datasets, we use the `petastorm.reader.Reader` class, which implements the iterator interface that allows us to use plain Python to go over the samples very efficiently. The `petastorm.reader.Reader` class can be created using the `petastorm.make_reader` factory method:

```
from petastorm import make_reader
with make_reader('dfs://some_dataset') as reader:
    for sample in reader:
        print(sample.id)
        plt.imshow(sample.image1)
```

2. The following code example shows how we can stream a dataset into the TensorFlow `Examples` class, which as we have seen before is a named tuple with the keys being the ones specified in the Unischema of the dataset, and the values are `tf.tensor` objects:

```
from petastorm.tf_utils import tf_tensors
with Reader('dfs://some_dataset') as reader:
    tensor = tf_tensors(reader)
    with tf.Session() as sess:
        sample = sess.run(tensor)
        print(sample.id)
        plt.imshow(sample.image1)
```

In this way, we can read Parquet data into TensorFlow native data structures. In the future, they will be enabled to access the data using the `tf.data.Dataset` interface.

3. Similar to how we can load data into TensorFlow, data loaded using the Petastorm dataset can be incorporated into PyTorch using an adapter class, `petastorm.pytorch.DataLoader`, as is shown in the next Python code example:

```
from petastorm.pytorch import DataLoader
with DataLoader(Reader('dfs://some_dataset')) as train_
loader:
    sample = next(iter(train_loader))
```

```
    print(sample['id'])
    plt.plot(sample['image1'])
```

The fact that Petastorm stores the data in the Parquet data format, which is natively supported by Spark, has the advantage that it enables us to use a wide range of PySpark methods to analyze and handle the dataset.

4. The next Python code example shows you how to read a Petastorm dataset as a Spark DataFrame instance:

```
from petastorm.spark_utils import dataset_as_rdd
rdd = dataset_as_rdd('dfs://some_dataset', spark,
    [first_schema.id, first_schema.image1])
print(rdd.first().id)
```

Now that our Petastorm dataset has been loaded as a Spark DataFrame, it allows us to use standard tools to read and visualize it. Consider that this data has not been decoded and it will be useful to see the values of fields that have data types that are meaningful to display, such as, for example, strings or scalars.

5. For example, in the next code, we show how we can, given the path of our dataset, create a Spark DataFrame object from a Parquet file:

```
dataframe = spark.read.parquet(dataset_url)
```

6. Then we can show a schema as with any Spark DataFrame:

```
dataframe.printSchema()
```

7. We can also perform operations such as `count`:

```
dataframe.count()
And do operations like show on a single column:
dataframe.select('id').show()
We can alto utilize SQL as it can be used to query a
Petastorm dataset:
number_of_rows = spark.sql(
    'SELECT count(id) '
    'from parquet.`file:///example_dataset`').collect()
```

Once we can load the Petastorm dataset, we can then use it to prepare data for distributed deep learning training, as we will exemplify in the next section.

Using Petastorm to prepare data for deep learning

This section will demonstrate a workflow on Azure Databricks to use Spark to load and preprocess data, and save data using the Parquet format in the optimized FUSE mount dbfs:/ml location. Then we will load the data using Petastorm and pass it into a deep learning framework for training or inference:

1. First, we will create a unique working directory in the dbfs:/ml FUSE mount, which is optimized for machine and deep learning applications:

    ```
    import os
    import subprocess
    import uuid
    work_dir = os.path.join("/ml/tmp/petastorm",
                            str(uuid.uuid4()))
    dbutils.fs.mkdirs(work_dir)
    def get_local_path(dbfs_path):
      return os.path.join("/dbfs", dbfs_path.lstrip("/"))
    ```

    Although it is possible to use many file formats in PySpark, in this example we will use the MNIST dataset in LIBSVM format and load it using Spark's built-in LIBSVM data source.

2. LIBSVM is a commonly used open source machine learning library that is optimized for use for kernelized support vector machines. Each LIBSVM file is a text format file where each line represents a labeled sparse feature vector using the following format:

    ```
    label index1:value1 index2:value2
    ```

3. Where the indices are one-based and in ascending order. After loading, the feature indices are converted to be zero-based. Our first step is to download the data, so in the next code example, we will get the data from a URL and load it as a Spark DataFrame, specifying the number of features in the file, which in our case is 784:

    ```
    data_url = "https://www.csie.ntu.edu.tw/~cjlin/
    libsvmtools/datasets/multiclass/mnist.bz2"
    libsvm_path = os.path.join(work_dir, "mnist.bz2")
    subprocess.check_output(["wget", data_url, "-O", get_
    local_path(libsvm_path)])
    ```

```
df = spark.read.format("libsvm") \
    .option("numFeatures", "784") \
    .load(libsvm_path)
```

4. The Petastorm library allows us to work with scalar and array columns in the Spark DataFrame where we load our data. Here, our data is represented using MLlib vectors, which is a user-defined type that requires special handling. We will first need to register a user-defined function that will transform MLlib vectors into dense arrays. We can do this by running the following Scala code in a cell in our Azure Databricks notebook:

```
%scala
import org.apache.spark.ml.linalg.Vector
val toArray = udf {v: Vector => v.toArray }
spark.sqlContext.udf.register("toArray", toArray)
```

5. We will use this user-defined function to transform MLlib sparse vectors into dense arrays that we will then write as files in the Parquet format. The Petastorm library will sample groups of rows of the Parquet file into batches. As mentioned previously, the batch size is an important parameter for the utilization of I/O and computing performance and will yield better results if the optimized FUSE mount in the dfs:/ml directory is used. The parameter that controls the batch size is the parquet.block.size option, which is passed to the parser:

```
parquet_path = os.path.join(work_dir, "parquet")
df.selectExpr("toArray(features) AS features",
              "int(label) AS label") \
    .repartition(10) \
    .write.mode("overwrite") \
    .option("parquet.block.size", 1024 * 1024) \
    .parquet(parquet_path)
```

We can use the Petastorm library to load data in the Parquet format to create, for example, a tf.data.Dataset class instance that we can then feed into our deep learning framework for training our models. Here, we will demonstrate how we can do this by creating a simple convolutional neural network model using Keras, which is a deep learning framework included in TensorFlow.

6. We will make all the necessary library imports and create a function that will retrieve the model with the desired architecture, which in our case will be a two-dimension convolutional layer followed by max-pooling and a dropout layer:

```
import tensorflow as tf
from tensorflow import keras
from tensorflow.keras import models, layers
from petastorm import make_batch_reader
from petastorm.tf_utils import make_petastorm_dataset
def get_model():
 model = models.Sequential()
 model.add(layers.Conv2D(32, kernel_size=(3, 3),
                           activation='relu',
                           input_shape=(28, 28, 1)))
 model.add(layers.Conv2D(64, (3, 3),
            activation='relu'))
 model.add(layers.MaxPooling2D(pool_size=(2, 2)))
 model.add(layers.Dropout(0.25))
 model.add(layers.Flatten())
 model.add(layers.Dense(128, activation='relu'))
 model.add(layers.Dropout(0.5))
 model.add(layers.Dense(10, activation='softmax'))
 return model
```

7. In versions of Azure Databricks Machine Learning Runtime before 7.0, we needed to whitelist _* files that were created when we saved data in the Parquet format, but now, the Azure Databricks Machine Learning Runtime has already pre-installed the pyarrow library, which automatically ignores all _* files. We can now define the Petastorm path of our dataset using the get_local_path function:

```
petastorm_dataset_url = "file://" + get_local_
path(parquet_path)
```

8. In the next Python code example, we will use the Petastorm make_batch_reader function to load groups of rows of the Parquet file in batches of data to train a Keras model that is obtained using the previously defined function:

```
with make_batch_reader(petastorm_dataset_url, num_
epochs=100) as reader:
```

```
dataset = make_petastorm_dataset(reader) \
.map(lambda x: (tf.reshape(x.features, [-1, 28, 28, 1]),
tf.one_hot(x.label, 10)))
model = get_model()
optimizer = keras.optimizers.Adadelta()
model.compile(optimizer=optimizer,
              loss='categorical_crossentropy',
              metrics=['accuracy'])
model.fit(dataset, steps_per_epoch=10, epochs=10)
```

9. Here, we can see the output of the training process in an Azure Databricks notebook running ML Runtime:

```
Cmd 16

1  with make_batch_reader(petastorm_dataset_url, num_epochs=100) as reader:
2      dataset = make_petastorm_dataset(reader) \
3      .map(lambda x: (tf.reshape(x.features, [-1, 28, 28, 1]), tf.one_hot(x.label, 10)))
4      model = get_model()
5      optimizer = keras.optimizers.Adadelta()
6      model.compile(optimizer=optimizer,
7                  loss='categorical_crossentropy',
8                  metrics=['accuracy'])
9      model.fit(dataset, steps_per_epoch=10, epochs=10)
10


s removed, simply drop this attribute
  column_as_pandas = column.data.chunks[0].to_pandas()
10/10 [==============================] - 1s 149ms/step - loss: 49.8006 - accuracy: 0.0885
Epoch 2/10
10/10 [==============================] - 1s 122ms/step - loss: 46.9740 - accuracy: 0.1024
Epoch 3/10
10/10 [==============================] - 1s 116ms/step - loss: 46.6869 - accuracy: 0.0794
Epoch 4/10
10/10 [==============================] - 1s 125ms/step - loss: 45.8081 - accuracy: 0.0927
Epoch 5/10
10/10 [==============================] - 1s 134ms/step - loss: 43.4356 - accuracy: 0.0897
Epoch 6/10
10/10 [==============================] - 1s 125ms/step - loss: 40.5104 - accuracy: 0.1079
Epoch 7/10
10/10 [==============================] - 1s 122ms/step - loss: 39.6522 - accuracy: 0.0952
Epoch 8/10
10/10 [==============================] - 1s 121ms/step - loss: 38.5819 - accuracy: 0.0915
Epoch 9/10
10/10 [==============================] - 1s 126ms/step - loss: 38.0751 - accuracy: 0.1042
Epoch 10/10
10/10 [==============================] - 1s 113ms/step - loss: 37.3983 - accuracy: 0.1063

Command took 15.50 seconds -- by
```

Fig.9.3 – Training a model using Petastorm

In this example, we have defined a Keras deep learning model and trained it using a Parquet file that is read into batches of data to further optimize the process. As shown in this section, the Petastorm library provides us with efficient methods to use Parquet, a format natively supported in Spark and Azure Databricks, to train deep learning models. In the next section, we will dive into a concrete example of how we can train deep learning models for featurization using the Azure Databricks Machine Learning Runtime.

Data preprocessing and featurization

Featurization is the process that we use to transform unstructured data such as text, images, or time-series data into numerical continuous features that are more easily handled by machine and deep learning models. It can be differentiated from featuring engineering from the fact that in featuring engineering the variables are already in the numerical form or have a more defined structure that leads us to the need to refactor or transform these variables into something that makes the machine or deep learning algorithm easier to extract patterns. In featurization, we need to first define a way in which we will extract numerical features from the unstructured data that we have.

We have the need to perform featurization basically because our deep learning models cannot interpret unstructured data directly and therefore, we need not only to extract it but to do this in a computationally efficient manner. This process needs to be incorporated into our deep learning pipeline if we are working with models that need to be somehow frequently re-trained.

Azure Databricks Machine Learning Runtime provides us with support for deep learning featurization models that can be used to extract the before-mentioned features from our data. We have available pre-trained models that can be used to compute features that are used in other downstream processes or models. These pre-trained models can be created using the already included deep learning frameworks TensorFlow and PyTorch, which are already included in the Azure Databricks Machine Learning Runtime.

In this section, we will exemplify how we can perform the featurization of data in Azure Databricks by doing transfer learning, which is a deep learning method to reuse domain knowledge from one problem to another using featurization as the underlying technique. This featurization will be applied using pre-trained models to extract the numerical variables from the original data representation.

Featurization using a pre-trained model for transfer learning

In this section, we will demonstrate how we can use pre-trained models in the Azure Databricks Machine Learning Runtime to produce features that can be used for downstream models and processes. Here, we will show this by using these extracted features for transfer learning to be used to train a new model.

As mentioned earlier, featurization is a technique that is used for feature encoding to be able to apply the domain knowledge of one model to a different but related problem. We can see this from the pragmatic point of view of using the patterns extracted to perform one task to be able to extrapolate this to a new task, with the potential of both improving the efficiency and reducing the amount of time necessary for training a new model.

This section will demonstrate how you can extract features for transfer learning using a pre-trained TensorFlow model using the Azure Databricks Machine Learning Runtime, using the following workflow:

1. We will start with a pre-trained deep learning model – in this case, an image classification model from `tensorflow.keras.applications`.

2. We will remove the last layer of the pre-trained deep learning model for it to produce a tensor of numerical features as output, rather than the actual prediction of the model.

3. We will apply this model to a different domain knowledge image dataset to be able to use it to encode numerical features from it.

4. We can then use these encoded features to train a new downstream model, although we will omit this final step in this example.

In this example, we will use the `pandas` library **User Defined Functions** (**UDFs**) to apply, in a distributed manner, the featurization to our dataset. The pandas UDF is an efficient method to apply this featurization as part of a transformation process of our dataset and as well as the `Scalar Iterator`, pandas UDFs provide us with flexible and high-performance support for deep learning applications.

As mentioned before, in this section we will load our dataset and extract features from it using pandas UDFs. The steps that we will take are to load the data using into and Spark DataFrame reading from the binary data source, loading the pre-trained model that will be used for transfer learning, and finally, applying it to encode features from the dataset using Scalar Iterator pandas UDFs.

Featurization using pandas UDFs

The workflow that is used to encode new features using pandas UDFs using pre-trained models can be generalized as follows:

1. First, we need to load our data into a DataFrame.

2. Afterward, we need to prepare the pre-trained model that will be used for featurization.

3. Next, we will define the image loading process and how we will apply the feature encoding.

4. Finally, we extract the features by applying the pre-trained model as a pandas UDF.

This workflow is a generalization and can also be applied to image preprocessing and custom model training pipelines. It is also efficient since it takes advantage of pandas UDFs to improve the performance of the feature extraction:

1. Our first step is to do the necessary library imports, which include instantiating the ResNet50 pre-trained model, which is included in the TensorFlow framework:

   ```
   import pandas as pd
   from PIL import Image
   import numpy as np
   import io
   import tensorflow as tf
   from tensorflow.keras.applications.resnet50 import
   ResNet50, preprocess_input
   from tensorflow.keras.preprocessing.image import img_to_
   array
   from pyspark.sql.functions import col, pandas_udf,
   PandasUDFType
   ```

2. We will use for this example the flowers dataset as our example dataset. This dataset contains flower photos stored under five folders, each of them representing one type of flower. This dataset is also stored in the directory of the Azure Databricks dataset examples directory:

   ```
   %fs ls /databricks-datasets/flower_photos
   ```

3. After we have identified the path to our dataset, we can then load images using the Spark API binary file as a data source. We could alternatively use the native Spark API image data source, but the binary file data source will provide us with more flexibility in the way in which we process the images:

```
images = spark.read.format("binaryFile") \
    .option("pathGlobFilter", "*.jpg") \
    .option("recursiveFileLookup", "true") \
    .load("/databricks-datasets/flower_photos")
```

4. After the data has been loaded, we can visualize it using the display function and limit the result to 5 records:

```
display(images.limit(5))
```

5. Here, we see the output of this process:

```
Cmd 17

1   import pandas as pd
2   from PIL import Image
3   import numpy as np
4   import io
5   import tensorflow as tf
6   from tensorflow.keras.applications.resnet50 import ResNet50, preprocess_input
7   from tensorflow.keras.preprocessing.image import img_to_array
8   from pyspark.sql.functions import col, pandas_udf, PandasUDFType
9
10  images = spark.read.format("binaryFile") \
11      .option("pathGlobFilter", "*.jpg") \
12      .option("recursiveFileLookup", "true") \
13      .load("/databricks-datasets/flower_photos")
14
15  display(images.limit(5))
```

▶ (1) Spark Jobs

▶ ▦ images: pyspark.sql.dataframe.DataFrame = [path: string, modificationTime: timestamp ... 2 more fields]

	path	modificationTime	length	content
1	dbfs:/databricks-datasets/flower_photos/tulips/2431737309_1468526f8b.jpg	2019-12-11T22:18:32.000+0000	281953	/9j/4AAQSkZJRgABAQEBLAEsAAD/4gxYSUNDX1BST0ZJTEUAAQEAAAxiT (truncated)
2	dbfs:/databricks-datasets/flower_photos/sunflowers/4932735362_6e1017140f.jpg	2019-12-11T22:18:00.000+0000	277326	/9j/4AAQSkZJRgABAQEASABIAAD/2wBDAAEBAQEBAQEBAQEBAQECAgf (truncated)
3	dbfs:/databricks-datasets/flower_photos/tulips/8717900362_2aa508e9e5.jpg	2019-12-11T22:18:52.000+0000	265806	/9j/4AAQSkZJRgABAQEASABIAAD/4gxYSUNDX1BST0ZJTEUAAQEAAAxiT (truncated)
4	dbfs:/databricks-datasets/flower_photos/sunflowers/4341530649_c17bbc5d01.jpg	2019-12-11T22:17:56.000+0000	257418	/9j/4AAQSkZJRgABAQEASABIAAD/4gxYSUNDX1BST0ZJTEUAAQEAAAxiT (truncated)

Showing all 5 rows

Fig.9.4 – Binary image files

6. Next, we will prepare the pre-trained model that we will use for featurization by downloading it and truncating the last layer. In this example, we will use the ResNet50 pre-trained model as discussed before. When loading the model, there are some concepts that are good to have in mind that concern the size of the models. If we are dealing with a moderate-sized model of less than 1 GB, it is good practice to download the model to the cluster driver and then use it to broadcast the weights to the workers. This is the approach that will be used in this example. The alternative approach is when we have models larger than 1 GB. In these cases, it is better to load the model weights directly to the distributed storage.

7. We will instantiate our model and specify removing the last layer, display the model summary to verify that the last layer was removed, and then finally broadcast the model weights to the workers:

```
model = ResNet50(include_top=False)
model.summary()  # verify that the top layer is removed
bc_model_weights = sc.broadcast(model.get_weights())
```

8. The next step is to define a function that returns a ResNet50 model with the last layer truncated with the pretrained weights broadcast to the workers:

```
def model_fn():
   model = ResNet50(weights=None, include_top=False)
   model.set_weights(bc_model_weights.value)
   return model
```

Now that we have our function to instantiate the truncated and broadcast model, we can define the pandas UDF that will be used to featurize a pd.Series array of images using Scalar Iterator pandas UDFs to improve the performance of loading large models on the workers.

9. The next function will be used to preprocesses raw image bytes and will be later used for prediction:

```
def preprocess(content):
   img = Image.open(io.BytesIO(content)).resize([224,
224])
   arr = img_to_array(img)
   return preprocess_input(arr)
```

10. Next, we define another function, which in turn will be the one in charge of featurizing a pd.Series array of raw images using the previously defined function to decode them, applying the pre-trained model to it, and returning a pd.Series of image features. Here, we have flattened the output multidimensional tensors to better handle them when loading them to a Spark DataFrame:

```
def featurize_series(model, content_series):
    input = np.stack(content_series.map(preprocess))
    preds = model.predict(input)
    output = [p.flatten() for p in preds]
    return pd.Series(output)
```

11. The next step is to define a Scalar Iterator pandas UDF wrapping for our featurization function. The decorator used here will specify that the output of this process will be the Spark DataFrame column of type ArrayType(FloatType). The content_series_iter argument specifies to iterate over batches of data, where each batch is a pandas Series array of images. Using the Scalar Iterator pandas UDF allows us to load the model once and use it several times over different batches of data, improving the performance of the process of loading big models:

```
@pandas_udf('array<float>', PandasUDFType.SCALAR_ITER)
def featurize_udf(content_series_iter):
    model = model_fn()
    for content_series in content_series_iter:
        yield featurize_series(model, content_series)
```

This way we have successfully registered our function that later can be used to obtain our features based on our pre-trained model.

Applying featurization to the DataFrame of images

The pandas UDF allows us to manage a large set of unstructured data records, which could be, for example, large images but can give rise to problems such as **Out of Memory (OOM)** errors:

1. We can give a hit that allows such errors in the cell below, to be handled in a more efficient way by reducing the arrow batch size using the maxRecordsPerBatch option:

```
spark.conf.set("spark.sql.execution.arrow.
maxRecordsPerBatch", "1024")
```

2. We can now run featurization over our entire Spark DataFrame. Consider that this operation can take a considerable amount of time since it applies a large model to the full dataset:

```
features_df = images.repartition(16).select(col("path"),
featurize_udf("content").alias("features"))

features_df.write.mode("overwrite").parquet("dbfs:/ml/
tmp/flower_photos_features")
```

3. Here, we see the process running, which can take some time to dump the results as Parquet files:

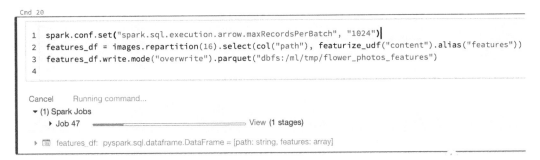

Fig.9.5 – Extracting and writing features

Finally, we have computed the extracted features using the pre-trained model and now we are able to use them to train a new model using the domain knowledge extracted from the pre-trained model. This last step is not shown in this example as it is considered a different process.

Summary

In this chapter, we discussed how we can use TFRecords and Petastorm as libraries to make the process of loading a large amount of data easier to train our distributed deep learning models. This led to us learning how these records are structured, how we can handle expensive operations such as automated schema inference, how we can prepare records to be consumed, and how we can use them not only in the context of deep learning frameworks but also for pure Python applications.

We finished the chapter with an example of how we can leverage having pre-trained models to extract new features based on domain knowledge that later can be applied to extract features to train a new model.

In the next chapter, we will learn how we can fine-tune the parameters of our deep learning models to improve their performance in Azure Databricks.

10
Model Tracking and Tuning in Azure Databricks

In the previous chapter, we learned how to create machine learning and deep learning models, as well as how to load datasets during distributed training in Azure Databricks. Finding the right machine learning algorithm to solve a problem using machine learning is one thing, but finding the best hyperparameters is another equally or more complex task. In this chapter, we will focus on model tuning, deployment, and control by using MLflow as a Model Repository. We will also use Hyperopt to search for the best set of hyperparameters for our models. We will implement the use of these libraries using deep learning models that have been made using the scikit-learn Python library.

More concretely, we will learn how to track runs of the machine learning model's training to find the most optimal set of hyperparameters, deploy and manage version control for the models using MLflow, and learn how to use Hyperopt as one of the alternatives to operations such as random and grid search for model tuning. We will be covering the following topics:

- Tuning hyperparameters using AutoML
- Automating model tracking with MLflow

- Hyperparameter tuning with Hyperopt
- Optimizing model selection with scikit-learn, Hyperopt, and MLflow

Before we dive into these concepts, let's go through the requirements for this chapter.

Technical requirements

To work on the examples in this chapter, you must have the following:

- An Azure Databricks subscription.
- An Azure Databricks notebook attached to a running cluster with Databricks Runtime ML version 7.0 or higher.

Tuning hyperparameters with AutoML

In machine learning and deep learning, hyperparameter tuning is the process in which we select a set of optimal hyperparameters that will be used by our learning algorithm. Here, hyperparameters are values that are used to control the learning process. In contrast, other parameters will be learned from the data. In this sense, a hyperparameter is a concept that follows its statistical meaning; that is, it's a parameter from a prior distribution that captures the prior belief before we start to learn from the data.

In machine learning and deep learning, it is also common to call hyperparameters the parameters that are set before we start to train our model. These parameters will control the training process. Some examples of hyperparameters that are used in deep learning are as follows:

- Learning rate
- Number of epochs
- Hidden layers
- Hidden units
- Activation functions

These parameters will directly influence the performance and training time of our models, and their selection plays a crucial role in the success of our model. For example, a neural network with a learning rate that's too low will fail to accurately capture patterns in the observed data. Finding good hyperparameters requires us to efficiently map the search space if possible, which can be a daunting task. This is because to find each set of good values, we need to train a new model, and this is an operation that can be expensive in terms of time and computational resources. Some of the techniques that are used for this include common algorithms such as Grid Search, Random Search, and Bayesian Optimization.

One of the techniques that we can apply to search for the best hyperparameters more efficiently is **AutoML,** which stands for **automated machine learning**. This is the process of automatically applying machine learning to solve optimization problems, which in our case is to optimize the hyperparameters that are used in our training algorithm. This helps us overcome problems that might arise from manually searching for the right hyperparameters or by applying techniques such as Grid and Random Search, which are algorithms that must be run for a long time. This is because they search for all the possible values in the search space without evaluating how promising these areas might be.

Azure Databricks Runtime for Machine Learning (Databricks Runtime ML) provides us with two options to automatically map the search space of possible hyperparameters. These are known as MLflow and Hyperopt, two open source libraries that apply AutoML to automate model selection and hyperparameter tuning.

HyperOpt is a Python library designed for large-scale Bayesian optimization for the hyperparameters of models. It can be used to scale the search process to several computing cores in a distributed manner. Although we will just focus on the hyperparameter optimization side of things, it can be also used to optimize pipelines, as well as for data preprocessing, learning algorithm selection, and, of course, hyperparameter tuning. To use HyperOpt, we need to define an optimizer that will be applied to the desired function to optimize any function, which in our case can be to maximize the performance metric or to minimize the loss function. In our examples, HyperOpt will take a search space of hyperparameters and will move according to the previous results, thus moving over the search space in an informed manner, differentiating itself from algorithms such as random search and grid search.

The other available library in the Azure Databricks Runtime for Machine Learning that's used to automatically map the search space of possible hyperparameters is MLflow, which we have discussed in previous chapters. MLflow is an open source platform that is used to manage end-to-end machine learning and deep learning model life cycles and supports automated model tuning tracking. This automated model tracking comes from its integration with the Spark MLlib library. It allows us to track which hyperparameters yield the best results by using `CrossValidator` and `TrainValidatorSplit`, which automatically log the validation metrics and hyperparameters to make it easier to get the best models.

In the upcoming sections, we will learn how to apply both HyperOpt and MLflow to track the obtain the best hyperparameters for the models that have been trained with the Azure Databricks Runtime for Machine Learning.

Automating model tracking with MLflow

As we mentioned previously, MLflow is an open-source platform for managing machine and deep learning model life cycles, which allows us to perform experiments, ensure reproducibility, and support easy model deployment. It also provides us with a centralized model registry. As a general overview, the components of MLflow are as follows:

- MLflow Tracking: It records all data associated with an experiment, such as code, data, configuration, and results.

- MLflow Projects: It wraps the code in a format that ensures the results can be reproduced between runs, regardless of the platform.

- MLflow Models: This provides us with a deployment platform for our machine learning and deep learning models.

- Model Registry: The central repository for our machine learning and deep learning models.

In this section, we will focus on MLflow Tracking, which is the component that allows us to log and register the code, properties, hyperparameters, artifacts, and other components related to training deep learning and machine learning models. The MLflow Tracking component relies on two concepts, known as experiments and runs. The experiment is where we execute the training process and is the primary unit of organization; all runs belong to an experiment. Therefore, these experiments refer to a specific run and we can visualize it, compare it, and download logs and artifacts related to it. The following information is stored in each MLflow:

- Source: This is the notebook that the experiment was run in.

- Version: The notebook version or Git commit hash if the run was triggered from an MLflow Project.

- Start and end time: The start and end time of the training process.

- Parameters: A dictionary containing the model parameters that were used in the training process.

- Metrics: A dictionary containing the model evaluation metrics. MLflow records and lets you visualize the performance of the model throughout the course of the run.

- Tags: These run metadata saved as key-value pairs. You can update tags during and after a run completes. Both keys and values are strings.

- Artifacts: These are any data files in any format that are used in the run, such as the model itself, the training data, or any other kind of data that's used.

In Azure Databricks Runtime for Machine Learning, every time we use CrossValidator or TrainValidationSplit in the code while training our algorithm, MLflow will store the hyperparameters and evaluating metrics to make visualizing and, ultimately, finding the optimal model, easier.

In the following sections, we will learn how to use CrossValidator and TrainValidationSplit while training our algorithm to leverage the advantages of MLflow and visualize the hyperparameters that yielded the best results.

Managing MLflow runs

When we use CrossValidator or TrainValidationSplit while training the model, we will have nested MLflow runs. These will be nested as follows:

- Main run: The information for CrossValidator or TrainValidationSplit is logged to the main run. If there is no active run, MLflow will create a new run and log into it, ending it before exiting the process.

- Child runs: Each hyperparameter value and its corresponding performance metrics are logged in a child run that is dependent on the parent run.

These nested runs are common when we are performing a hyperparameter search for multi-step processing. For example, we can have nested runs that look as follows:

```
with mlflow.start_run(nested=True):
  mlflow.log_param("lr", 0.05)
  mlflow.log_param("batch_size", 256)
  with mlflow.start_run(nested=True):
    mlflow.log_param("max_runs", 16)
    mlflow.log_param("epochs", 10)
    mlflow.log_metric("rmse", 98)
  mlflow.end_run()
```

These nested runs will be shown in the MLflow UI as a tree that can be expanded so that we can view the results in more detail. This allows us to keep the results from different child runs organized in the main run.

When discussing MLflow run management, we need to be sure that we are logging into the main run. To ensure this, when the `fit()` function is called, this should be wrapped inside a `mlflow.start_run()` statement to log the information into the right run. As we have seen already, this way, we can easily place metrics and parameters inside the run. If the `fit()` function is called many times in the same active run, MLflow will append unique identifiers to the names of the parameters and metrics that are used to avoid any possible conflicts in the naming.

In the next section, we will see an example of how to use MLflow tracking with MLlib to find the optimal set of hyperparameters for a machine learning model.

Automating MLflow tracking with MLlib

In this section, we will exemplify the use of MLflow for tracking the performance of a PySpark MLlib `DecisionTreeClassifier` model. MLflow will keep track of the model's learning and will also allow us to store all the artifacts used in the process. We will center our attention on examining the hyperparameters that yielded the best results to find the optimal settings. The model will be trained using the MNIST dataset, which is included in the Databricks example datasets. The dataset is in LIBVSM format and is divided into train and test data. It contains two columns – one for the label and the other for the image encoded in 784 features. Let's get started:

1. First, we will load the data while specifying the number of features and cache the data in the worker memory:

```
train_data = spark.read.format("libsvm") \
  .option("numFeatures", "784") \
    .load("/databricks-datasets/mnist-digits/data-001/
mnist-digits-train.txt")
test_data = spark.read.format("libsvm") \
  .option("numFeatures", "784") \
    .load("/databricks-datasets/mnist-digits/data-001/
mnist-digits-test.txt")
train_data.cache()
test_data.cache()
```

2. After this, we can display the number of records in each dataset and display the training data:

```
print("Train images:{} ; test images {}".format(train_
data.count(), test_data.count()))
display(train_data)
```

To pass these features to the machine learning model, we need to do some feature engineering. We can standardize these operations into a single pipeline using MLlib, which allows us to pack several preprocessing operations into a single workflow. MLib Pipelines is very similar to the pipeline implementation in scikit-learn and has five main components. These are as follows:

- DataFrame: The actual DataFrame that holds our data.

- Transformer: The algorithm that will transform the data in a features DataFrame into data for a predictions DataFrame, for example.

- Estimator: This is the algorithm that fits on the data to produce a model.

- Parameter: The multiple `Transformer` and `Estimators` parameters specified together.

- Pipeline: The multiple Transformer and Estimators operations combined into a single workflow.

Here, we will have a pipeline composed of two steps, which are a `StringIndexer` to transform the labels from numerical into categorical features, and the `DecisionTreeClassifier` model, which will predict the label based on the train data in the feature's column. Let's get started:

3. First, let's make the necessary imports and create the pipeline:

```
from pyspark.ml.classification import
DecisionTreeClassifier, DecisionTreeClassificationModel
from pyspark.ml.feature import StringIndexer
from pyspark.ml import Pipeline
```

4. Now, we can instantiate StringIndexer and DecissionTreeClassifier:

```
indexer = StringIndexer(inputCol="label",
                        outputCol="indexLabel")
model = DecisionTreeClassifier(labelCol="indexLabel")
```

5. Finally, we can chain StringIndexer and DecissionTreeClassifier together into a single workflow:

```
pipeline = Pipeline(stages=[indexer, model])
```

6. So far, we have followed the standard way of creating a normal pipeline. What we will do now is include the CrossValidator MLflow class so that we can run the cross-validation process of the model. The evaluation metrics of each validated model will be tracked by MLflow and will allow us to investigate which hyperparameters yielded the best results.

   In this example, we will specify two hyperparameters in the CrossValidator MLflow class to be examined, which are as follows:

   a) `maxDepth`: This parameter determines the maximum depth that the tree can grow in the DecissionTreeClassifier. Deeper trees can yield better results but are more expensive to train and returns models that tend to overfit more.

b) `maxBins`: This parameter determines the number of bins that will be generated to discretize the continuous features into a finite set of numbers. It is a parameter that trains the model in a more efficient way when the training is done in a distributed computing environment. In this example, we will start by specifying a value of two, which will convert the grayscale features into either 1 or 0s. We will also test with a value of 4 so that we have greater granularity.

7. Now, we can define the evaluator that we will be using to measure the performance of our model. We will use PySpark `MulticlassClassificationEvaluator` and then use `weightedPrecision` as a metric:

```
from pyspark.ml.evaluation import
MulticlassClassificationEvaluator

model_evaluator =
MulticlassClassificationEvaluator(labelCol="indexLabel",
metricName="weightedPrecision")
```

8. Next, we will define the grid of parameters we want to examine:

```
from pyspark.ml.tuning import CrossValidator,
ParamGridBuilder
hyperparam_grid = ParamGridBuilder() \
    .addGrid(model.maxDepth, [2, 6]) \
    .addGrid(model.maxBins, [2, 4]) \
    .build()
```

9. Now, we are ready to create CrossValidator using the previously defined pipeline, evaluator, and hyperparameter grid of values. As we mentioned previously, CrossValidator will keep track of the models we've created, as well as the hyperparameters we've used:

```
cross_validator = CrossValidator(
    estimator=pipeline,
    evaluator=model_evaluator,
    estimatorParamMaps=hyperparam_grid,
    numFolds=3)
```

10. Once we have defined the cross-validation process, we can start the training process. If an MLflow tracking server is available, it will start to log the data from each run that we do, along with all the other artifacts being used in the run under the current active run. If we don't have an active run, it will create a new one:

```
import mlflow
import mlflow.spark
with mlflow.start_run():
  cv_model = cross_validator.fit(train_data)
  test_metric = model_evaluator.evaluate(cv_model.
transform(test_data))
  mlflow.log_metric(f'model_metric_{model_evaluator.
getMetricName()}', test_metric)
  mlflow.spark.log_model(spark_model=cv_model.bestModel,
artifact_path='best-model')
```

11. In the preceding code, we created a new run to track the model. Remember that we are using the `with mlflow.start_run():` statement to avoid running into any naming conflicts if we run the cell multiple times. The preceding steps will guarantee that the `cv.fit()` function returns the best model that it was able to find, after which we evaluate its performance on the test data and then log the results of this evaluation in MLflow:

Figure 10.1 – Tracking a training run using MLflow

12. Once we have completed the training process, we will be able to see the results on the MLflow UI in the Azure Databricks Workspace, under the **Models** tab. We can also see the results of the training process by clicking on the *experiment* icon in the notebook where we are training our model. We can easily compare results from different runs by clicking on **Experiment Runs**, which will allow us to view all the notebook runs:

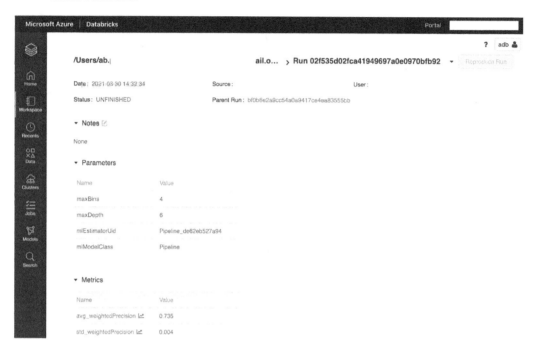

Figure 10.2 – The MLflow Model Registry UI

We can explicitly look for the specific results of these runs by passing a parameter in the **search run box**. For example, we can search the results we obtained when we used the value of 6 in the maxDepth hyperparameter by passing params. maxDepth = 6 in the search box.

We can also compare results by creating scatter plots of the different values of the hyperparameters that were used against a specific performance metric, which is useful when we're trying to find and compare the best set of hyperparameters.

The next section will dive into a more automatic approach of finding the best set of hyperparameters that minimize a certain function. We will do this using the loss function. We will be using Hyperopt to define and explore a search space and find the hyperparameters that minimize the loss function.

Hyperparameter tuning with Hyperopt

The Azure Databricks Runtime for Machine Learning includes Hyperopt, a Python library that is intended to be used on distributed computing systems to facilitate the learning process for an optimal set of hyperparameters. At its core, it's a library that receives a function that we need to either minimize or maximize, and a set of parameters that define the search space. With this information, Hyperopt will explore the search space to find the optimal set of hyperparameters. Hyperopt uses a stochastic search algorithm to explore the search space, which is much more efficient than using a traditional deterministic approach such as random search or grid search.

Hyperopt is optimized for use in distributed computing environments and provides support for libraries such as PySpark MLlib and Horovord, the latter of which is a library for distributed deep learning training that we will focus on later. It can also be applied in single-machine environments and with other more common libraries such as TensorFlow and PyTorch.

The general workflow when using Hyperopt is as follows:

1. First, we must find the function that we want to optimize. This can be either a performance metric that we want to maximize or a loss function that we want to minimize.

2. Next, we must define the conditional search space in which Hyperopt will scan for the best parameters to optimize the defined target function. Hyperopt supports several algorithms being used in the same run, which allows us to define more than one model to explore.

3. After that, we must define the search algorithm that will be used to scan the search space.

4. Finally, we can start the exploration process by running the `fmin()` Hyperopt function. This will take the objective function as an argument and the search space for identifying the optimal set of parameters.

This being said, the two most crucial parts of Hyperopt are as follows:

-   The objective function that we will pass as the optimization target. Generally, this is a loss function or a performance metric.

-   A dictionary of hyperparameters that define the search space. We can choose different distributions for each set of hyperparameter values.

In the next section, we will learn how to apply Hyperopt to the Azure Databricks Runtime for Machine Learning to find the best hyperparameters for a certain task, regardless of the algorithm that's applied.

Hyperopt concepts

In this section, we will go through certain concepts of Hyperopt to identify its core functionalities, as well as how we can use them to efficiently scan the search space for hyperparameters that help minimize or maximize our objective function. As we mentioned previously, when we start using Hyperopt, we must define the following:

- The objective function

- The search space that we will scan

- The database that will be used to save the result of evaluating the points in the search space

- The search algorithm that will be used to scan the search space

Here, our main goal is to find the best set of scalars that represent the optimal hyperparameters. Hyperopt allows us to define the search space in detail to provide information regarding where in the search space the target function will perform better. Other optimization libraries regularly assume that the search space is a subset of a larger vector space with the same distribution, which is not usually the case. The way Hyperopt makes us define the search space allows us to perform a more efficient scan.

Hyperopt allows us to define the objective function very simply, but it also has an incremental complexity when we increase the number of parameters, we want to keep track of during our execution. The simplest example is as follows, where we find the value of x that minimizes a linear function, $y(x) = x$:

```python
from hyperopt import fmin, tpe, hp
best_param = fmin(
    fn=lambda x: x,
    space=hp.uniform('x', 0, 1),
    algo=tpe.suggest,
    max_evals=100)
print(best_param)
```

We can decompose the preceding execution by breaking down its parameters:

- The Hyperopt `fmin()` function takes in the `fn` keyword argument and the function to minimize. In this case, we are using a very simple lambda function; that is, `f(y)=x`.

- The space argument is used to pass the search space. Here, we are using the `hp` function from Hyperopt to specify a uniform search space for the x variable in the range of 0 to 1. `hp.uniform` is a function that's built into the library that helps us specify the search space for specific variables.

- The `algo` parameter is used to select the search algorithm. In this case, `tpe` stands for Parzen estimators. We can also set these parameters to `hyperopt.random.suggest`. The search algorithm is an entire subject, so we will only briefly mention this, but you are free to search for more details about the types of algorithms that Hyperopt applies to find the most suitable set of hyperparameters for you.

- Lastly, the maximum number of evaluations is specified using the `max_evals` parameters.

Once we have specified all the required parameters, we can start the run, which will return a Python dictionary of values for the specified variable that minimizes the objective function.

Even though our minimum working example here looks very simple, Hyperopt has a complexity that increases with the number of specifications that we have when optimizing an objective function. Some of the questions that we can ask ourselves when we're coming up with the best strategy are as follows:

- What type of information is required besides the actual returned value of the `fmin()` function? We can obtain statistics and other data that is collected during the process and might be useful for us to keep.

- Is it necessary for us to use an optimization algorithm that requires more than one input value?

- Should we parallelize this process and if so, do we want communication between these parallel processes?

Answering these questions will define the way in which we can implement the optimization strategy.

In a parallel optimization process, communication occurs between the Hyperopt optimization algorithm and the objective function to obtain the actual value of the objective function in that specific point of the search space. In this communication, the value that returns the objective function in that floating-point loss, which is also called a negative utility, is associated with that point in the search space.

The simple example shown here exemplifies how simple it is to define an objective function and a search space using Hyperopt. However, it still falls short in cases where we want to store more information than just the floating-point value that's returned by the objective function in the search space. It is also an evaluation that doesn't interact with the search algorithm or the concurrent function.

To tackle the first problem, we can use an objective function that retrieves more than one value by returning a nested dictionary with key values for the statistics and diagnostics that are desired. For this, we have a set of mandatory values that our objective function must return:

- `status`: We can choose one of the keys from `hyperopt.STATUS_STRINGS` to show the status of competition. This way, we can track if our evaluation was completed successfully or if it failed.

- `loss`: This is the float value that is returned by the actual objective function when it is evaluated in that specific point in the search space. It must be present if the status of the competition is OK.

We can also specify some additional optional arguments in the option of the objective function. These key values are as follows:

- `loss_variance`: A float value specifying the certainty of the stochastic objective function.

- `true_loss`: You can store the loss of the model with this name so that we can use the built-in plotting tool. This is especially useful when we are working with hyperparameter optimization because it allows us to plot the results of the exploration very simply.

- `true_loss_variance`: The uncertainty of the loss of the model.

This functionality allows us to store these values in databases that are JSON-compatible, such as MongoDB. The following code example shows us how we can define a very simple $f(x) = x$ objective function that returns the status of the execution as a key valued dictionary:

```
import pickle
import time
```

```
from hyperopt import fmin, tpe, hp, STATUS_OK
def objective(x):
    return {'loss': x, 'status': STATUS_OK }
best = fmin(objective,
    space=hp.uniform('x', -5, 5),
    algo=tpe.suggest,
    max_evals=50)
print(best)
```

We can take full advantage of being able to return dictionaries by using the `Trials` object, which allows us to store much more information while executing the optimization. In the following example, we are modifying the objective function to return a more complex output that will, later, be stored in the `Trials` object. We can do this by passing it as an argument of the Hyperopt `fmin()` function:

```
import pickle
import time
from hyperopt import fmin, tpe, hp, STATUS_OK, Trials
def objective(x):
    return {
        'loss': x,
        'status': STATUS_OK,
        'eval_time': time.time(),
        'more_data': {'type': None, 'value': [0, 1]},
        'attachments':
            {'time_module': pickle.dumps(time.time)}
        }
trials_obj = Trials()
best_params = fmin(objective,
    space=hp.uniform('x', -5, 5),
    algo=tpe.suggest,
    max_evals=50,
    trials=trials_obj)
print(best_params)
```

Once the execution has been completed, we can access the `Trials` object to inspect the values that were returned by the objective function during the optimization. We have different way we can access the values stored in the `Trials` object:

- `trials.trials`: Returns a list of dictionaries with all the parameters of the search.
- `trials.results`: The dictionaries that are returned by the objective function.
- `trials.losses()`: The actual losses of all the successful trials.
- `trials.statuses()`: The status of each of the trials.

This `Trials` object can be saved as a pickle object or parsed using custom code so that we have a deeper understanding of the results of the trials.

The attachments of the trials can be accessed like so:

```
msg = trials.trial_attachments(trials.trials[0])['time_module']
time_module = pickle.loads(msg)
```

In this example, we are fetching the `time_module` attachment of the first trial.

One thing that is important to note is that in Azure Databricks, we have the option to pass both the `Trials` and `SparkTrials` objects to the trials parameter of the `fmin()` function. The SparkTrials object is passed when we execute single-machine algorithms such as scikit-learn models as objective functions. Here, the Trials object is used in distributed training algorithms such as MLlib or Horovod models. The SparkTrials class allows us to distribute the execution of the optimization without having to introduce any custom logic. It also improves the distribution of the computation to the workers.

Do not use the SparkTrials object with algorithms designed for distributed training as that will cause the execution to be parallelized in the cluster.

Now that we have a clearer picture of how we can configure the basic parameters of the `fmin()` Hyperopt function, we can dive into how the search space is defined.

Defining a search space

We can express the search space as a set of nested functions that define the individual search space for different test cases. For example, in the following code block, we are specifying the search space for the x parameter with two test spaces called `test_space_1` and `test_space_2`:

```
from hyperopt import hp
```

```
search_space = hp.choice('a',
    [('test_space_1', 1 + hp.lognormal('x', 0, 1)),
    ('test_space_2', hp.uniform('y', -5, 5))
    ])
```

Here, Hyperopt will sample from these nested stochastic expressions according to the search algorithm. This algorithm applies an adaptative exploration strategy rather than just taking samples from the search space. In the preceding example, we defined three parameters:

- a: Selects the case to be used
- test_space_1: Generates positive values for the x parameter
- test_space_2: Generates positive values for the y parameter

Each expression has the label as the first argument, which is used internally by the algorithm to return the parameter choice to the caller. In this example, the code works like so:

- If the a variable is 0, we will use x and not y.
- If the a variable is 1, then y will be used but not x.

The x and y variables are conditional parameters that depend on the result of the value of a. If we encode the variables in this conditional way, we are stating that the x variable has no effect on the objective function when a is 0, which helps the search algorithm assign credit in a more efficient way. This takes advantage of the domain knowledge that the user has on the objective function to be analyzed.

search_space is a variable space that's defined as a graph expression that describes how to sample a point without generating the search space. This optimizes both the memory and computational use of the execution. These graph expressions are called pyll graphs and are defined in the hyperopt.pyll class.

You can explore a sample search space by just sampling it:

```
import hyperopt.pyll.stochastic
print(hyperopt.pyll.stochastic.sample(search_space))
```

We know now how we can define the search space where we will optimize our objective function. It is important to note that the way we define the cases to be tested in the search space will also impact the way the search algorithm will work on the trials during the adaptive search. In the next section, we will learn about some of the best practices for using Hyperopt in Azure Databricks.

Applying best practices in Hyperopt

So far, we have discussed how we can specify both the execution of the objective function to be minimized and the search space that we will evaluate it in. Also, we have seen that there is plenty of flexibility in terms of how we can define the way we will explore the possible values of the hyperparameters, as well as how we can take advantage of the domain knowledge by defining conditional parameters in the search space.

We must keep some aspects of the Hyperopt library itself in mind, as well as the way it is executed in Azure Databricks. Here are some concepts that are important to remember while optimizing objective functions with Hyperopt in Azure Databricks:

- The Hyperopt Tree of Partzen Estimator algorithm is a Bayesian method that's much more efficient than common grid and random search approaches. It allows us to scan a larger set of possible hyperparameters. Apply the domain knowledge when defining the search space to improve the performance of the exploration.

- When we use `hyperopt.choice()` and MLflow to track the progress of the optimization, Mlflow will log the index of the choice list. We can fetch the parameter values using the Hyperopt `hyperopt.space_eval()` function.

- When working on large datasets, it is always advisable to experiment on small subsets of the data to incrementally learn about the optimal search space. This helps us define the one that will be used on the entire dataset.

- When we use the SparkTrials object, it is advisable to use CPU-only clusters. This is because in Azure Databricks, parallelism is reduced in the GPU clusters in comparison to the CPU ones.

- Do not use SparkTrials on autoscaling clusters as the parallelism value is selected at the beginning of the execution. Therefore, if the cluster scales, it won't improve the performance of the execution.

Applying these concepts will ensure that we take advantage of the execution of the optimization with Hyperopt in Azure Databricks. In the next section, we will learn more about how to improve the inference of deep learning models in Azure Databricks.

Optimizing model selection with scikit-learn, Hyperopt, and MLflow

As we saw in the previous sections, Hyperopt is a Python library that allows us to track optimization runs that can be used for hyperparameter model tuning distributed computing environments such as Azure Databricks. In this section, we will go through an example of training a scikit-learn model. We will use Hyperopt to track the tuning process and log the results to MLflow, the model life cycle management platform.

In Azure Databricks Runtime for Machine Learning, we have an optimized version of Hyperopt at our disposal that supports MLflow tracking. Here, we can use the SparkTrials objects to log the results of the tuning process of single-machine models during parallel executions. We will use these tools to find the best set of hyperparameters for several scikit-learn models.

We will do the following:

- Prepare the training dataset.
- Use Hyperopt to define the objective function to be minimized.
- Define an Hyperopt search space, over which we will scan for the best hyperparameters.
- Define the search algorithm that will be used to scan the search space.
- Execute the optimization and track the best set of parameters in MLflow.

Let's get started:

1. First, we will import all the necessary libraries that we will be using:

```
import numpy as np
from sklearn.datasets import fetch_california_housing
from sklearn.model_selection import cross_val_score
from sklearn.svm import SVC
from sklearn.ensemble import RandomForestClassifier
from sklearn.linear_model import LogisticRegression
from hyperopt import fmin, tpe, hp, SparkTrials, STATUS_
OK, Trials
import mlflow
```

2. In this example, we will be using the California housing dataset, which is included as an example dataset in scikit-learn. This dataset holds data from the 1990 US census with house values for 20,000 houses in California, along with some features with data about income, number of inhabitants, rooms, and so on. To fetch this dataset, we will use the `fetch_california_housing()` scikit-learn function and dump the results into variables named features and target:

```
features, target = fetch_california_housing(return_X_
y=True)
```

3. Now that we have split our dataset into the features and the target, we will scale the prediction variables, which is a common practice in machine learning and deep learning. This way, we will ensure we have scaling consistency. This will help us overcome several problems that arise when our features have not been discretized into scaled variables. The features that are available in our dataset are as follows:

a) Median house income per block

b) House age

c) Average number of rooms in a house per block

d) Average number of bedrooms

e) Total block population

f) Average number of house occupants

g) Latitude

h) Longitude

4. By reviewing the scale of the features in each column – a simple operation that can be done in a pandas DataFrame using the `describe()` method – we can see that the orders of magnitude in the mean values vary a lot:

```
import pandas as pd
features_df = pd.DataFrame(features)
features_df.describe()
```

5. As we can see, the ranges and scales of these variables can be very different:

Cmd 2

```
1  import pandas as pd
2
3  features_df = pd.DataFrame(features)
4  features_df.describe()
```

Out[5]:

	0	1	2	3	4	5	6	7
count	20640.000000	20640.000000	20640.000000	20640.000000	20640.000000	20640.000000	20640.000000	20640.000000
mean	3.870671	28.639486	5.429000	1.096675	1425.476744	3.070655	35.631861	-119.569704
std	1.899822	12.585558	2.474173	0.473911	1132.462122	10.386050	2.135952	2.003532
min	0.499900	1.000000	0.846154	0.333333	3.000000	0.692308	32.540000	-124.350000
25%	2.563400	18.000000	4.440716	1.006079	787.000000	2.429741	33.930000	-121.800000
50%	3.534800	29.000000	5.229129	1.048780	1166.000000	2.818116	34.260000	-118.490000
75%	4.743250	37.000000	6.052381	1.099526	1725.000000	3.282261	37.710000	-118.010000
max	15.000100	52.000000	141.909091	34.066667	35682.000000	1243.333333	41.950000	-114.310000

Command took 0.17 seconds -- by   14:10:55 on dplearn

Figure 10.3 – The statistical summary of the features

6. The population of a block is measured in thousands, but the average number of bedrooms is centered around 1.1. To prevent situations in which features with large values are deemed to be more important, scaling the features and normalizing them is a standard practice. To normalize the features so that they're the same scale, we can use the scikit-learn `StandardScaler` function:

```
from sklearn.preprocessing import StandardScaler
scaler = StandardScaler()
scaled_features = scaler.fit_transform(features)
```

7. Now that we have scaled the predictor features, we can check that the mean is close to 0 by calling the `mean(axis=0)` NumPy method on the `scaled_features` dataset:

```
scaled_features.mean(axis=0)
```

8. Now that our predictor features have been normalized and scaled, we will have to convert the numeric target column into discrete values:

Cmd 3

```
1  from sklearn.preprocessing import StandardScaler
2  scaler = StandardScaler()
3  scaled_features = scaler.fit_transform(features)
4  print(scaled_features.mean(axis=0))
```

```
[ 6.60969987e-17  5.50808322e-18  6.60969987e-17 -1.06030602e-16
 -1.10161664e-17  3.44255201e-18 -1.07958431e-15 -8.52651283e-15]
```

Command took 0.03 seconds -- by   at 30/03/2021, 14:11:48 on dplearn

Figure 10.4 – The scaled feature means

9. This target column vector is the value of each house and is a continuous positive scalar. We will transform this into a target column vector with discrete values that are 1 when the house value is greater than the mean and 0 where it is not. This way, the prediction can be framed as a question; for example, given the predictor features, is the values of this house greater than the mean? To discretize the values, we will compare this with the mean and use the numpy.where() function to convert the Booleans values that the comparison yields into 1s and 0s:

```
target_discrete = np.where(target < np.median(target), 0,
1)
```

This way, we have encoded our target variable into a column vector with two levels, either 1 or 0. This indicates whether the housing price is above the mean.

10. Now that we have prepared our dataset, we can start setting the stage for optimizing the hyperparameters of several scikit-learn models that we will use to train an effective classification model. To this end, we will use the following three scikit-learn classifier models:

a) Support Vector Machines

b) Logistic Regression

c) Random Forest

11. We will create an objective function to be minimized that will be passed to Hyperopt. This will run the training and calculate the cross-validation performance metric over several types of models, defined as 'type' parameters. One thing to bear in mind about this function is that because we are using the `fmin()` Hyperopt function, this will try to minimize the objective function. So, to improve the performance metric – which is accuracy, in this case – we must use the negative accuracy. Otherwise, Hyperopt will try to minimize the accuracy, which is something that we obviously don't want to do:

```python
def objective(params):
    classifier_type = params['type']
    del params['type']
    if classifier_type == 'svm':
        clf = SVC(**params)
    elif classifier_type == 'rf':
        clf = RandomForestClassifier(**params)
    elif classifier_type == 'logreg':
        clf = LogisticRegression(**params)
    else:
        return 0
    accuracy = cross_val_score(
                        clf,
                        scaled_features,
                        target_discrete).mean()
    return {'loss': -accuracy, 'status': STATUS_OK}
```

Notice that the output of this function is a key-valued dictionary with the two required parameters. These are the 'loss', which is the negative accuracy, and the 'status', which is defined using one of the Hyperopt status strings that we imported previously.

12. The next step is to define the search space we will use to scan all the possible hyperparameters for each of the selected scikit-learn models. This will help us find the one that yields the best accuracy. We will use the `hyperopt.choice()` function to select the different models and specify the search space for each classification algorithm. The label of this selection is `classifier_type`. It will iterate over the previously defined types of algorithms in the objective function based on the given labels in the search space:

```
search_space = hp.choice('classifier_type', [
    {
        'type': 'svm',
        'C': hp.lognormal('SVM_C', 0, 1.0),
        'kernel': hp.choice('kernel', ['linear', 'rbf'])
    },
    {
        'type': 'rf',
        'max_depth': hp.quniform('max_depth', 1, 3, 5),
        'criterion': hp.choice('criterion', ['gini',
'entropy'])
    },
    {
        'type': 'logreg',
        'C': hp.lognormal('LR_C', 0, 1.0),
        'solver': hp.choice('solver', ['liblinear',
'lbfgs'])
    },
])
```

The `hyperopt.choice()` function that is inside the search space expression will pass the `'type'` parameter to the objective function, which, in turn, creates an internal variable named `classifier_type` and uses it to select the appropriate algorithm. It then passes the remaining parameters to the model. Notice how, in the objective function, the `'type'` parameter was removed after the internal variable was created to avoid any possible naming conflicts in the target algorithm.

13. Now, we must choose the search algorithm that will be used to scan the search space. As we have seen previously, Hyperopt provides us with two main options for this:

a) `hyperopt.tpe.suggest`: This is the **Tree of Parzen Estimators** (**TPE**) algorithm, which is a Bayesian algorithm that adaptively selects hyperparameters based on the previous results.

b) `hyperopt.rand.suggest`: This is a random search algorithm that differs from the TPE algorithm in that it is a non-adaptive approach that uses a sampling strategy over the search space.

14. In this example, we will use the TPE algorithm, so we will choose `tpe.suggest`, as shown here:

```
search_algorithm = tpe.suggest
```

15. The next step is to define the object that will hold the results of each of the trials. We will use SparkTrials to keep track of the progress of the exploration and log it in MLflow. As we mentioned previously, we will use the SparkTrials object instead of the native `Trials` Hyperopt object because we are using algorithms designed to be run on a single machine. If we were to use an MLlib algorithm, we must use the MLflow API to keep track of the `Trials` result.

In turn, SparkTrials is an object that can receive two optional arguments, as follows:

a) `parallelism`: This is the number of algorithms to fit and evaluate in parallel. The default value of this parameter is equal to the number of available Spark task slots.

b) `timeout`: Maximum time (in seconds) that `fmin()` can run for. There is no maximum time limit.

The default `parallelism` value in Spark is 8. The recommended setting for this is to explicitly set this value to a positive value, since the Spark task slots is dependent on the cluster's size.

16. The final step of preparing the optimization run is to use the `mlflow.start_run()` MLflow wrapper. We will use it to log the results of this exploration in the MLflow platform. This will automatically keep track of the progress of the run, along with all the defined variables and labels in the objective function and search space:

```
with mlflow.start_run():
    best_hyperparams = fmin(
```

```
    fn=objective,
    space=search_space,
    algo=search_algorithm,
    max_evals=32,
    trials= SparkTrials())
```

17. The `max_evals` parameter allows us to specify the maximum number of points in the search space that will be evaluated. So, we can consider this value to be the maximum number of models that Hyperopt will fit and evaluate for each search space:

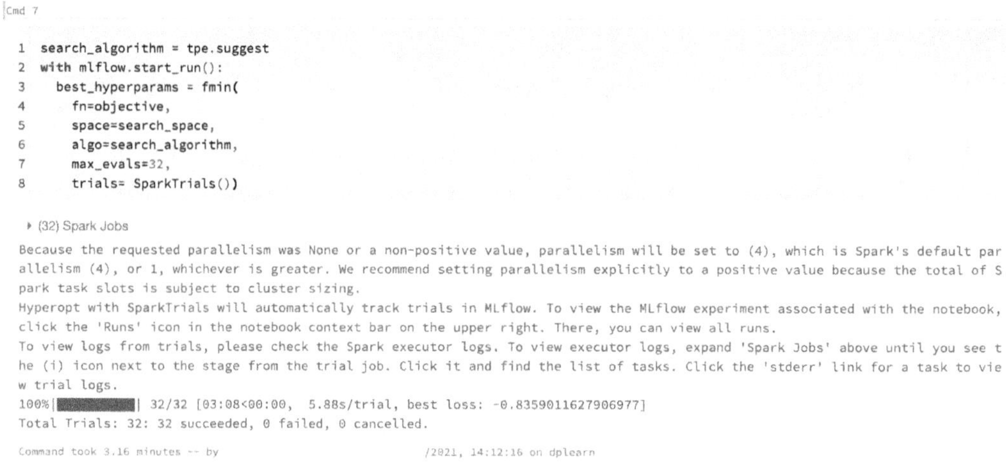

```
Cmd 7

1  search_algorithm = tpe.suggest
2  with mlflow.start_run():
3      best_hyperparams = fmin(
4          fn=objective,
5          space=search_space,
6          algo=search_algorithm,
7          max_evals=32,
8          trials= SparkTrials())
```

▸ (32) Spark Jobs

Because the requested parallelism was None or a non-positive value, parallelism will be set to (4), which is Spark's default parallelism (4), or 1, whichever is greater. We recommend setting parallelism explicitly to a positive value because the total of Spark task slots is subject to cluster sizing.
Hyperopt with SparkTrials will automatically track trials in MLflow. To view the MLflow experiment associated with the notebook, click the 'Runs' icon in the notebook context bar on the upper right. There, you can view all runs.
To view logs from trials, please check the Spark executor logs. To view executor logs, expand 'Spark Jobs' above until you see the (i) icon next to the stage from the trial job. Click it and find the list of tasks. Click the 'stderr' link for a task to view trial logs.
100%|██████████| 32/32 [03:08<00:00, 5.88s/trial, best loss: -0.8359011627906977]
Total Trials: 32: 32 succeeded, 0 failed, 0 cancelled.

Command took 3.16 minutes -- by   /2021, 14:12:16 on dplearn

Figure 10.5 – Tracking the training run with MLflow

18. Once the execution has been finalized, we can examine the hyperparameters that produced the best result:

```
import hyperopt
print(hyperopt.space_eval(search_space,
                          best_hyperparams))
```

We should see the best set of hyperparameters being printed in the console as a "key-value dictionary" along with the names and values of the best parameters found over the search space for the defined minimization function:

```
Cmd 8

1  import hyperopt
2  print(hyperopt.space_eval(search_space, best_hyperparams))


{'C': 3.6002403259280142, 'kernel': 'rbf', 'type': 'svm'}

Command took 0.02 seconds -- by                    at 30/03/2021, 14:12:24 on dplearn
```

Figure 10.6 – Best hyperparameters

We can view the MLflow experiment by clicking on the *experiment* icon in the top right of the notebook. This option will display all the results that were obtained during the runs and, as we mentioned previously, allow us to visualize the set of hyperparameters and algorithms that yielded the best results. We can also examine the effect of varying a specific parameter.

In this way, we can tune several models at the same time and keep track of the performance of the models in each of the defined search spaces. Thanks to Hyperopt and MLflow, we can easily define an advanced pipeline to fine-tune models in parallel using Azure Databricks Runtime for Machine Learning.

Summary

In this chapter, we learned about some of the valuable features of Azure Databricks that allow us to track training runs, as well as find the optimal set of hyperparameters of machine learning models, using the MLflow Model Registry. We have also learned how we can optimize how we scan the search space of optimal parameters using Hyperopt. This is a great set of tools because we can fine-tune models that have complete tracking for the hyperparameters that are used for training. We also explored a defined search space of hyperparameters using adaptive search strategies, which are much more optimized than the common grid and random search strategies.

In the next chapter, we will explore how to use the MLflow Model Registry, which is integrated into Azure Databricks. MLflow makes it easier to keep track of the entire life cycle of a machine learning model and all the associated parameters and artifacts used in the training process, but it also allows us to deploy these models easily so that we can make REST API requests to get predictions. This integration of MLflow in Azure Databricks enhances the entire workflow process of creating and deploying machine learning models.

11
Managing and Serving Models with MLflow and MLeap

In the previous chapter, we learned how we can fine-tune models created in Azure Databricks. The next step is how we can effectively keep track and make use of the models that we train. Software development has clear methodologies for keeping track of code, having stages such as staging or production versions of the code and general code lifecycle management processes, but it's not that common to see that applied to machine learning models. The reasons for this might vary, but one reason could be that the data science team follows its own methodologies that might be closer to academia than the production of software, as well as the fact that machine learning doesn't have clearly defined methodologies for development life cycles. We can apply some of the methodologies used commonly in software for machine learning models in Azure Databricks.

This chapter will focus on exploring how the models and processes we deploy can be used for inference. To do this, we have to provide a way to pack the used pipelines and work out how to deploy them to serve predictions. We will investigate this in more detail with the `MLflow` library, an open source platform integrated into Azure Databricks that allows us to manage the entire end-to-end machine learning model life cycles from training to deployment.

`MLflow` allows you to keep track of the machine learning training process. The training process is called an *experiment* and we can access previous experiments to record and compare parameters, centralize model storage, and manage the deployment of the model as a **REpresentational State Transfer (REST) application programming interface (API)**. We will learn how to use `MLflow` in combination with the tools learned in the previous chapters. The examples will focus on the tracking and logging of the training process of machine learning algorithms and on how to deploy and manage models and apply version control to them.

The concepts we will explore are listed as follows:

- Managing the machine learning models
- Model Registry example
- Exporting machine learning pipelines using MLeap
- Serving models with MLflow

Before we dive into these concepts, let's go through the technical requirements to execute the examples shown here.

Technical requirements

To work on the examples given in this chapter, you need to have the following:

- An Azure Databricks subscription
- An Azure Databricks notebook attached to a running cluster with Databricks Runtime for Machine Learning (Databricks Runtime ML) with version 7.0 or higher

Managing machine learning models

As we have seen before, in Azure Databricks we have at our disposal the MLflow Model Registry, which is an open source platform for managing the complete lifecycle of a machine learning or deep learning model. It allows us to directly manage models with a chronological linage, model versioning, and stage transition. It provides us with tools such as Experiments and Runs, which allow us to quickly visualize the results of training runs and hyperparameter optimization, and to maintain a proper model version control to keep track of which models we have available for serving and quickly update the current version if necessary.

MLflow has in Azure Databricks a Model Repository **user interface** (**UI**) in which we can set our models to respond to REST API requests for inference, transition models between stages, and visualize metrics and unstructured data associated with the models, such as description and comments. It gives us the possibility of managing our models either from the Model Registry API or directly from the Model Registry UI:

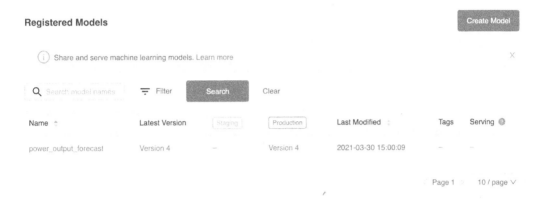

Figure 11.1 – Models in the MLflow registry

In this section, we will learn the core concepts of MLflow and reinforce some of the methods that we have seen in previous chapters.

Before we jump into the actual way in which we can manage models in MLflow, let's look at the core concepts associated with models, as follows:

- **Model**: This is the actual MLflow model that is logged in to the registry from an experiment or run using the mlflow.[model-flavor].log_model methods, after which it can then be registered in the Model Registry.

- **Registered Model** : This is a model that was first logged in to the Model Registry and then later registered in the model repository. This should have a unique name and has properties such as versions, lineage, and other associated artifacts.

- **Model Stage** : Models in MLflow can be in one or more of the available stages that we can assign to it. The available options of the stage are **None**, **Staging**, **Production**, and **Archived**. Any user that has the proper read and write permissions can transition a model between stages.

- **Model Version** : This is indicative of the number of times that a model was committed to the repository. This means that when we register a model it is assigned Version 1, and every time we register a model with the same name, the version will be incremented.

- **Activities**: Each of the events to which the model was subjected, such as stages transition, new versions being registered, and so on. This allows us to keep track of models in a more detailed way throughout all their life cycles.

- **Description**: Unstructured metadata that we can add to the models to have a better context of what is the purpose of the model, and any other relevant data surrounding its development or use.

Given this very summarized description of the basic concepts around `MLflow` Model Registry models in Azure Databricks, we can start to learn how we can register a model, transition it between stages, and make use of available models in the registry. These concepts will be shown in more detail in the next sections.

Using MLflow notebook experiments

As we have seen in previous sections, using `MLflow mlflow.start_run()` in an Azure Databricks notebook logs the metrics and parameters used in the run as an active experiment. Each experiment has a run ID, which is a unique numerical identifier. We can visualize a notebook experiment and its corresponding runs by clicking on the **Experiment** button in the notebook toolbar. This will show us the run parameters and associated metrics. We also have the option to download all the associated artifacts used in the run.

We cannot delete a notebook experiment because they are part of the notebook itself. This is important to note because if we use the MLflow `Client.tracking.delete_experiment()`, it will delete the notebook itself too.

All MLflow runs will logged into into the active experiment. If there is no active experiment, Azure Databricks will create a new one. We can set the active experiment in which we want to log our artifacts in the following ways:

- Using `mlflow.set_experiment()` to specify the experiment in which we want to log the run

- Setting the experiment ID as a parameter of the `mlflow.start_run()` command

- Setting environment variables named `MLFLOW_EXPERIMENT_NAME` or `MLFLOW_EXPERIMENT_ID`

If we don't specify any experiment, the run will be logged to the notebook experiment.

It is useful to note that we have two types of experiment in MLflow, outlined as follows:

- **Workspace experiment**: This is not associated with any notebook. Any job or notebook can log a run into one of them by specifying the workspace experiment ID.
- **Notebook experiment**: This is associated with a specific notebook and it's the default option when no experiment ID was passed as a parameter of the run. Only `MLflow` runs within that specific notebook can be logged into the Notebook experiment.

Workspace experiments can be created in the Azure Databricks Workspace UI or by using the `MLflow` API:

- We can use the runs of the experiments to query the metrics and parameters used on them. For example, if we want to query the runs in an experiment with an **R-squared (R2)** metric greater than 0.3, we can go to the Search Runs field in the Experiment tab in the Notebook toolbar and enter the following expression:

```
metrics.r2 > 0.3
```

- In a similar way, we can filter the runs that have a certain set of parameters using the following syntax:

```
params.max_depth = 4
params.max_depth = 4 AND metrics.accuracy > 0.8
```

- You also have the option to filter the runs based on the state or version of a model by using the Filter option in the Search Runs box .You can access drop-down menus with the State and Linked models.
- We can track the version of the Notebook or Git project used for a run in the following ways:

  a) We can click in the **Source** field of a run if this was created from an Azure Databricks notebook or job.

  b) If the run was created from a Git project, the **Source** field of the run will be a link to the branch of the Git project used to trigger the run.

- Finally, if it's a workspace experiment and the run was logged in to the experiment, the model will be displayed as an artifact that can be downloaded from the registered model page in the Model Registry UI.

Now that we have learned a bit more about the concepts of experiments and runs, we will learn in the next sections how to register and manage the lifecycle of a machine or deep learning model in Azure Databricks.

Registering a model using the MLflow API

As mentioned before, we have many ways in which we can register a model in the MLflow Model Registry. We can very simply use the MLflow API to register a model by using the mlflow.[model-flavor].log_model method to log and register a model in the registry. If no other model has been registered with the same name it will return a ModelVersion MLflow object and will name this first model as Version 1. If we have another model with the same name, it will increase the model version and return the new current version of the model:

- An example of this is shown in the following code snippet, using the Python MLflow API in Azure Databricks:

```
with mlflow.start_run(run_name=<run-name>) as run:
  <train and evaluate the model>
  mlflow.<model-flavor>.log_model(<model-flavor>=<model>,
    artifact_path="<model-path>",
    registered_model_name="<model-name>"
  )
```

- We can also register a model with a specified model name once the training in the experiment is over. This is especially useful if we have developed several models and we want to register one of them specifically in the registry. We can do this using the mlflow.register_model() method and passing the model path as an argument. We can do this in the following way:

```
result_model=mlflow.register_model(
"runs:<new_model_path>",
"<model-to-update-name>")
```

If we already have a model with the same name, this will update the version of the model; otherwise, it will create a new model with a name of Version 1.

- Also, we can use the MLflow create_registered_model() method to create a new registered model, as illustrated in the following code snippet. Although this approach is similar to the previous method, this will raise an MlflowException error if the model is in the Model Registry already:

```
client = MlflowClient()
result = client.create_registered_model("<model-name>")
```

Now that we have the model registered in the MLflow Model Registry, we can control the stage of the model. We will see how to do this in the next section.

Transitioning a model stage

We have mentioned before that we have several stages that can be assigned to a model. Specifically, these stages are None, Staging, Production, and Archived. As is common in other data engineering environments, the Staging stage is where all the models that are in an experimental or testing phase can be found, and the Production stage is where all the models that are already tested and are now being consumed by production processes are to be found. Finally, the Archived stage is where we place all the models that have been replaced by newer versions or that are scheduled to be deleted. The stages of models are not exclusive, so a model can belong to more than one stage.

Any user with the proper permissions can request a model to be transitioned to another stage, but if you have full permissions, you can do this transition with no intermediary. If you are falling short on the permissions on that model, the request can be reviewed by someone with higher access to it.

You can see the requests regarding transitioning models in the **Model Registry UI** in the following screenshot:

Figure 11.2 – Request to move a model to another stage

We will get started with the transition as follows:

1. To transition a model between stages, we can use either the **Model Registry** UI or the MLflow API to do this programmatically. To do this, we can use the transition_model_version_stage() method, as illustrated in the following code snippet:

```
client = MlflowClient()
client.transition_model_version_stage(
    name="<our_model>",
    version=<model-version>,
    stage="<target_stage>",
    description="<description>"
)
```

As mentioned before, we can pass the next possible values in the stage argument of the `transition_model_version_stage()` method:

a) `staging`

b) `archived`

c) `production`

d) `none`

2. The description parameter allows us to pass any relevant information regarding that model. For example, you can include what the purpose of the development of that model was or which kind of approach was taken when it was built, and why. We can just update the description of the model by using the MLflow `update_model_version()` method, as we can see in the following code example:

```
mlflow_client = MlflowClient()
mlflow_client.update_model_version(
    name="<model-name>",
    version=<model-version>,
    description="<updated_description>"
)
```

3. We can also just update the name of a registered model, using the MLflow `client.rename_registered_model()` method. Conceptually, we can do this in the following way:

```
mlflow_client=MlflowClient()
mlflow_client.rename_registered_model("<model-name>",
    "<new-name>")
```

4. If we are not sure about the model name and version that we want to update, we can always get a list of all registered models in the Model Registry by calling the MLflow client API `list_model_versions()` method, as is conceptually shown in the following code snippet:

```
from pprint import pprint
mlflow_client = MlflowClient()
all_models = [pprint(rm) for rm in mlflow_client.list_
registered_models()]
```

5. If you want to find the current version of a specified model name and fetch all the details, we can use the MLflow client `search_model_version()` method, as illustrated in the following code snippet:

```
from pprint import pprint
ml_client=MlflowClient()
[pprint(mv) for mv in client.search_model_
versions(f"name={model_name}")]
```

6. If we want to remove a specific model version, we can specify a list of versions that we want to delete and use the MLflow client `delete_model_version()` method. For example, if we wanted to delete versions 1 and 2 of the `house_price_classifier` model, we would run the following code:

```
client = MlflowClient()
versions = [1, 2]
for version in versions:
    client.delete_model_version(name="house_price_
classifier", version=version)
```

In this section, we have gone through some of the most common methods to manage the model lifecycle with MLflow. The next section will be a full working example in which we will train a simple model, register it on the Model Registry, and use it to do inference on a new observation once it has been deployed.

Model Registry example

In this section, we will go through an example in which we will develop a machine learning model and use the MLflow Model Registry to save it, manage the stages in which it belongs, and use it to make predictions. The model will be a Keras neural network, and we will use the Windfarm US dataset to predict the power output of wind farms based on parameters from weather conditions such as wind direction, speed, and air temperature. We will make use of MLflow to keep track of the stage of the model and be able to register and load it back again to make predictions:

1. First, we will retrieve the dataset that will be used to train the model. We will use the pandas `read_csv()` function to load directly from the **Uniform Resource Identifier (URI)** of the file in GitHub, as follows:

```
import pandas as pd
wind_farm_data = pd.read_csv("https://github.com/
dbczumar/model-registry-demo-notebook/raw/master/dataset/
windfarm_data.csv", index_col=0)
```

The dataset has predictive features such as the wind speed, direction, and air temperature taken every 8 hours, and as a target the total daily power output over a period of 7 years. Our objective will be to train a deep learning neural network to predict the power output of a farm given these predictive features.

2. After the data has been loaded into a Pandas DataFrame , we will split the data into training and validation data, as follows:

```
training_data = pd.DataFrame(wind_farm_data["2014-01-
01":"2018-01-01"])
training_data_x = training_data.drop(columns="power")
training_data_y = training_data["power"]
validation_data = pd.DataFrame(wind_farm_data["2018-01-
01":"2019-01-01"])
validation_x= validation_data.drop(columns="power")
validation_y = validation_data["power"]
```

3. The next step is to define a model that will be used to make predictions on new data. We will define a simple TensorFlow Keras model that we can use in the following way:

```
import tensorflow as tf
from tensorflow.keras.layers import Dense
from tensorflow.keras.models import Sequential
model = Sequential()
model.add(Dense(120,
          input_shape=( training_data_x.shape[-1],),
          activation="relu",
          name="hidden_layer"))
model.add(Dense(1))
```

4. We have defined a simple neural network with one single hidden layer of 120 neurons with a ReLU activation function. The output is a single dense layer of a single cumulative neuron. The next step is to compile the model, for which we will use **mean-squared error** (**MSE**) as the performance metric and Adam as optimizer, and fit it on the training data. To register the model into the MLflow Model Registry, we will use the MLflow with the mlflow.start_run() method and the mlflow.tensorflow.autolog() method to automatically capture all the model artifacts and parameters.

The code is shown in the following snippet:

```
import mlflow
import mlflow.keras
import mlflow.tensorflow
with mlflow.start_run():
  mlflow.tensorflow.autolog()
  model.compile(loss="mse", optimizer="adam")
  model.fit(training_data_x, training_data_y,
            epochs=100, batch_size=32,
            validation_split=.2)
  run_id = mlflow.active_run().info.run_id
```

5. This will start the training process, and you will see the results of each one of the epochs in the console, as follows:

```
> Cmd 15

1   import mlflow
2   import mlflow.keras
3   import mlflow.tensorflow
4   with mlflow.start_run():
5     mlflow.tensorflow.autolog()
6     model.compile(loss="mse", optimizer="adam")
7     model.fit(training_data_x, training_data_y, epochs=100, batch_size=32, validation_split=.2)
8     run_id = mlflow.active_run().info.run_id
9

Cancel  ·•· Running command...

Epoch 1/100
 1/37 [..............................] - ETA: 14s - loss: 13650072.0000WARNING:tensorflow:Callback
method `on_train_batch_end` is slow compared to the batch time (batch time: 0.0010s vs `on_train_ba
tch_end` time: 0.0038s). Check your callbacks.
37/37 [==============================] - 1s 17ms/step - loss: 10327475.6053 - val_loss: 7010075.500
0
Epoch 2/100
37/37 [==============================] - 0s 2ms/step - loss: 9103096.6579 - val_loss: 5787546.0000
Epoch 3/100
37/37 [==============================] - 0s 2ms/step - loss: 7767455.8816 - val_loss: 4857740.0000
Epoch 4/100
37/37 [==============================] - 0s 2ms/step - loss: 6206970.8947 - val_loss: 4506521.5000
Epoch 5/100
37/37 [==============================] - 0s 2ms/step - loss: 5468120.0921 - val_loss: 4548711.0000
Epoch 6/100
37/37 [==============================] - 0s 2ms/step - loss: 5595911.5132 - val_loss: 4584017.0000
Epoch 7/100
37/37 [==============================] - 0s 2ms/step - loss: 5188757.7697 - val_loss: 4589618.0000
Epoch 8/100
37/37 [==============================] - 0s 2ms/step - loss: 5723409.1711 - val_loss: 4563780.5000
Epoch 9/100
```

Figure 11.3 – Tracking the training of the model using MLflow

The model has now been logged in to the MLflow Model Registry, and all the related artifacts and performance metrics can be tracked there. We can now register the model using the MLflow API or the Model Registry UI in the Azure Databricks workspace:

1. To create a new registered model using the API, we can use the `mlflow.register_model()` function. This way, we will create a new model with a Version 1 name. We will specify a path where the model will be stored and a model name to do this, as follows:

```
model_path = "wind_farm_model"
model_name = "power_output_forecast"
model_uri = f"runs:/{run_id}/model"
model_details = \
mlflow.register_model(model_uri=model_uri,
                      name=model_name)
```

2. We can use MLflow not only to store the model artifacts but also to provide valuable context relating to the development of the model using descriptions and keeping lineage control, stating the model's version, as illustrated in the following code snippet:

```
from mlflow.tracking.client import MlflowClient
description="This model was trained as an example for the
Azure Databricks book with data from US wind farms to
predict the power output based on wind speed, direction
and air temperature."
mflow_client = MlflowClient()
mflow_client.update_registered_model(
  name=model_details.name,
  description= description
)
```

Providing meaningful descriptions helps to keep track of the purpose of the context of the development of the model, provide details about its attributes and usage, and—particularly—explicit the algorithm that was used and the data source used in the training. Descriptions then can be used to do the following:

a) Provide context that drives the development of the model.

b) Mention the methodology used and algorithm selected.

c) State differences between different versions of the same model.

d) Provide a high-level description of the data sources and feature engineering methods applied.

Now that we have properly described the model registered in MLflow, we can set it to one of the available stages. To do this, you will need to have the proper permissions at the user or model level. As mentioned before, we can choose to place our model in the None, Staging, Production, or Archived stage.

3. To transition a model between stages, we can use the MLflow client API `update_model_version()` method. If you don't have the appropriate level of permission, you can create a request for a model stage transition using the REST API. We will transition our model to the `Production` stage in the following code example:

```
mflow_client.transition_model_version_stage(
    name=model_details.name,
    version=model_details.version,
    stage='Production',
)
```

This results in the following output:

Cmd 17

```
1  from mlflow.tracking.client import MlflowClient
2
3  mflow_client = MlflowClient()
4
5  mflow_client.transition_model_version_stage(
6      name=model_details.name,
7      version=model_details.version,
8      stage='Production',
9  )
```

```
Out[30]: <ModelVersion: creation_timestamp=1617109202500, current_stage='Production', description
='', last_updated_timestamp=1617109209073, name='power_output_forecast', run_id='23d4e110709446b489
31ffe518cc6165', run_link='', source='dbfs:/databricks/mlflow-tracking/197254732634776/23d4e1107094
46b48931ffe518cc6165/artifacts/model', status='READY', status_message='', tags={}, user_id='7882912
336567795', version='4'>
```

Figure 11.4 – Transitioning the stage of a model

4. We can verify that the stage of the current model is Production, using the MLflow client API get_model_version() method, as follows:

```
model_vs = mflow_client.get_model_version(
  name=model_details.name,
  version=model_details.version,
)
print(f"The current model stage is: '{ model_vs.current_
stage }'"
```

5. The model is now in the Model Registry in the Production stage, so we can use it to forecast the power output of wind farms using new data. We will load the latest version of the model in the specified stage and use it to predict on the validation data, as follows:

```
from mlflow.tracking.client import MlflowClient
mflow_client = MlflowClient()
model_name = "power_output_forecast"
model_stage = 'Production'
model_version = mflow_client.get_latest_versions(model_
name, stages=[model_stage])[0].version
model_uri = f"models:/{model_name}/{model_stage}"
model = mlflow.pyfunc.load_model(model_uri)
power_predictions = pd.DataFrame(model.
predict(validation_x))
power_predictions.index = pd.to_datetime(weather_data.
index)
print(power_predictions)
```

The MLflow registry gives us the ability to have a lineage of the models in the different stages by keeping track of the version of the model. We can train and develop new models in the popular machine learning and deep learning frameworks such as scikit-learn and TensorFlow and in a very simple way deploy them to staging or production environments. Moreover, the MLflow library ensures that the application code keeps working properly once we transition the model into a different stage or create a new version.

In this example, we have very briefly shown how simple it is to create a new TensorFlow Keras model in Azure Databricks and manage its lifecycle using the MLflow Model Registry programmatically using Python.

In the next section, we will create a new version of the power prediction model using the popular scikit-learn framework, perform model testing in Staging, and update the production model by transitioning the new model version to Production.

Exporting and loading pipelines with MLeap

When we train a machine learning or deep learning model, our intention is to be able to use it several times to predict new observations of data. To do this, we must be able to not only store the model but also load it back again into one or more platforms. Therefore, we encounter the need to serialize the model for future use in scoring or predictions.

MLeap is a commonly used format to serialize and execute machine learning and deep learning pipelines made in popular frameworks such as Apache Spark, scikit-learn, and TensorFlow. It is commonly used for making individual predictions rather than batch predictions. These serialized pipelines are called bundles and can be exported as models and later be loaded and deployed back into Azure Databricks to make new predictions.

In this section, we will learn how to use MLeap to export and load back again a `DecisionTreeClassifier` MLlib model to make predictions using a saved pipeline in Azure Databricks. The steps that we will follow are listed here:

- We will create and fit a pipeline made using an MLlib model.
- We will serialize and export the pipeline using MLeap as a ZIP file and save it.
- We will load back the serialized model again and use it to make predictions on new unseen data.

It is important to note that MLeap is included with Azure Databricks Runtime ML, so if you are using a common runtime, you will need to install the `MLeap-Spark` package from the Maven repository at a cluster or notebook level, as well as the MLflow as a PyPi package to keep track of the experiment run and be able to also save the model in the MLflow Model Registry:

1. The first step will be to create the pipeline that we will serialize and reuse later. We will create an MLlib `DecisionTreeClassifier` that will be used to predict the labels of the News example dataset available in Azure Databricks. We will begin by first reading the data that will be used to train the model and cache it into the workers' memory, by running the following code:

```
training_data = spark.read.parquet("/databricks-datasets/
news20.binary/data-001/training").select("text", "topic")
training_data.cache()
```

2. After the data has been cached, we can visualize it and display the schema of the Spark data frame, as follows:

```
display(training_data)
training_data.printSchema()
```

3. The objective of our model will be to infer the topic of a new entry based on the training data that we have. You can see the first row of the schema in the following screenshot:

Figure 11.5 – Training data schema and first row

4. Once the data has been loaded, we can define our pipeline. We will have to apply feature engineering to transform the unstructured text data that we have as input into numerical features that can be consumed by the `DecisionTreeClassifier` MLlib model. To do this feature engineering, we will use the `StringIndexer`, `Tokenizer`, and `HashingTF` PySpark transformers to extract features from the text using a **term frequency-inverse document frequency** (**TF-IDF**) approach. We will begin to define the pipeline by defining the preprocessing functions to be used to extract the features, as follows:

```
from pyspark.ml.feature import StringIndexer, Tokenizer,
HashingTF
label_indexer = StringIndexer(
inputCol="topic",
```

```
    outputCol="label",
    handleInvalid="keep")
tokenizer = Tokenizer(inputCol="text",
                        outputCol="words")
hashing_tf = HashingTF(inputCol="words",
                        outputCol="features")
```

5. Next , we can define the model that will be used. As we mentioned before, we will use the MLlib `DecisionTreeClassifier` because of the support that it provides for multilabel classification. Here is the code we will use to define the model:

```
from pyspark.ml.classification import
DecisionTreeClassifier
classifier = DecisionTreeClassifier()
```

6. Now, we can very simply define the MLlib pipeline as the concatenation of the feature extraction operations and the classification model that we have selected, as follows:

```
from pyspark.ml import Pipeline
pipeline = Pipeline(stages=[label_indexer, tokenizer,
                            hashing_tf, dt])
```

7. The next step is to train and tune the pipeline that we have just defined. To do this, we will use the `CrossValidator` MLlib class to fine tune the number of features used in the TF-IDF process. In the following code block, we will define a parameter grid for the number of features extracted, and define the cross-validation to be run over our pipeline:

```
from pyspark.ml.evaluation import
MulticlassClassificationEvaluator
from pyspark.ml.tuning import CrossValidator,
ParamGridBuilder
hyperparam_grid = ParamGridBuilder().addGrid(
hashing_tf.numFeatures, [500, 750, 1000]).build()
cv = CrossValidator(estimator=pipeline,
evaluator=MulticlassClassificationEvaluator(),
estimatorParamMaps=hyperparam_grid)
```

8. Next, we will fit the `CrossValidator` over the training data and save the best model found for the specified search space of hyperparameters, as follows:

```
cv_model = cv.fit(training_data)
model = cv_model.bestModel
```

9. Finally, we can use the model to make predictions on the data, as illustrated in the following screenshot:

```
Cmd 6

1  cv_model = cv.fit(training_data)
2  model = cv_model.bestModel

▶ (60) Spark Jobs

MLlib will automatically track trials in MLflow. After your tuning fit() call has completed, view the MLflow UI to see logged runs.

Command took 3.59 minutes -- by          at 30/03/2021, 14:33:11 on dplearn
```

Figure 11.6 – MLflow logs show that the model is being tracked

10. Here, we will apply the model to the data that was used to train it, as follows:

```
predictions = model.transform(training_data)
display(predictions)
```

This results in the following output:

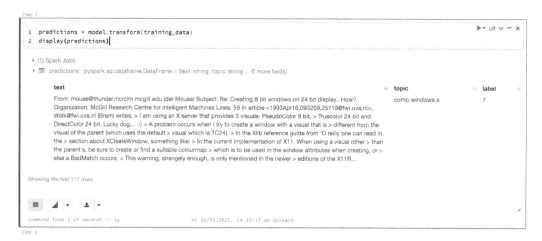

Figure 11.7 – Prediction of the topic of a text

So far, we have not introduced any major complexities, and our model is still just available in the current environment. We still need to transform this pipeline into a serialized bundle that can be exported to other environments to do predictions on new data.

11. We will use MLeap to serialize the model into a single ZIP file. We will create a directory where we will save our serialized model, as illustrated in the following code snippet:

```sh
%sh
mkdir /tmp/mleap__serialized_pipilene
```

12. Once the directories have been created, we can serialize the model using the `SimpleSparkSerializer` MLeap function, as follows:

```
import mleap.pyspark
from mleap.pyspark.spark_support import
SimpleSparkSerializer
model_path="/tmp/mleap__serialized_pipilene"
model_filename="serialized_pipeline.zip"
model_uri=f"jar:file:/{model_path}/{model_filename}"
model.serializeToBundle(model_uri, predictions)
```

Now, the MLlib pipeline has been serialized into a single bundle as a ZIP file. Always remember to check that the URI of the model starts with `jar:file` and ends with `zip` when the serialized bundle is a ZIP file.

13. As a next step, we will copy the file from the temporary directory into a location in the Databricks File System. In this way, we will simulate as is we have just fetched the serialized ZIP file bundle from a Model Registry such as MLflow or from a Blob or Data Lake storage. To copy the file from the temporary directory into the Databricks File System, we can use the `dbutils` tools, as shown in the following code snippet:

```
dbutils.fs.cp(f"file:/{model_path}/{model_filename}",
f"dbfs:/data/{model_name}")
display(dbutils.fs.ls("dbfs:/data"))
```

14. In the next part of the implementation, we will load and deserialize the pipeline. We can use this in a different execution context, thus giving us the capability to use the serialized pipeline to make new predictions. We can load back the pipeline that we have just serialized and saved in the Databricks File System, as illustrated in the following code snippet:

```
from pyspark.ml import PipelineModel
model_path="/tmp/mleap__serialized_pipilene"
model_filename="serialized_pipeline.zip"
```

```
model_uri=f"jar:file:{model_path}/{model_filename}"
deserialized_pipeline = PipelineModel.
deserializeFromBundle(model_uri)
```

We have loaded and deserialized the pipeline, and now it can be used to make predictions on unseen data. This is especially useful when we train our model in a Spark environment, but later this pipeline needs to be used in a non-Spark application. If you intend to use only Spark for making predictions, it is advisable to use the MLlib built-in methods for persisting and loading models.

15. Let's load some test data to make some predictions using the deserialized pipeline. We will load and display the test portion of the example News dataset, as follows:

```
test_data = spark.read.parquet("/databricks-datasets/
news20.binary/data-001/test").select("text", "topic")
test_data.cache()
display(test_data)
```

16. Now, we will use the deserialized pipeline to make predictions on the test data, as follows:

```
predictions = deserialized_pipeline.transform(test_data)
display(predictions)
```

17. As we can see in the following screenshot, we are able to obtain predictions by passing new data that has the same schema as the data used to fit the pipeline:

Figure 11.8 – Prediction of the loaded model

`MLeap` serialization is useful when we intend to run predictions outside of the Spark environment as it packs entire pipelines into serialized ZIP file format bundles. The applications running the predictions should be based in Scala or Java as `MLeap` has a format that is based on its own type of data frames, called `LeapFrames`. As a final concept, we can consider `MLeap` an alternative to the `MLflow` format for serializing machine learning models.

The next section will cover an example of how we can use `MLflow` to serve models on Azure Databricks.

Serving models with MLflow

One of the benefits of using MLflow in Azure Databricks as the repository of our machine learning models is that it allows us to simply serve predictions from the Model Registry as REST API endpoints. These endpoints are updated automatically on newer versions of the models in each one of the stages, therefore this is a complementary feature of keeping track of the model's lifecycle using the MLflow Model Registry.

Enabling a model to be served as a REST API endpoint can be done from the Model Registry UI in the Azure workspace. To enable a model to be served, go to the model page in the **Model Registry** UI and click on the **Enable Serving** button in the **Serving** tab.

Once you have clicked on the button, which is shown in the following screenshot, you should see the status as **Pending**. After a couple of minutes, the status will change to **Ready**:

Figure 11.9 – Enabling the serving of a model

If you want to disable a model for serving, you can click on the **Stop** button on the **Serving** tab on the page of the registered model.

To make a request to the deployed model, we need to know its unique URI. The URI of a deployed model depends also on the version of the model, so the resulting URI is something like this:

```
<databricks-instance>/model/<registered-model-name>/<model-version>/invocations
```

If we have a model called `house_classifier` and we want to know the URI of version 3 of this model, the code would look something like this:

```
https://<your_databricks-instance>/model/house_classifier/3/invocations
```

If this model has a particular stage assigned to it, it is also included in the URI of the endpoint. For example, the **Production** stage of the previous model will have an URI endpoint like this:

```
https://<your_databricks-instance>/model/house_classifier/Production/invocations
```

You can see a list of available URIs for the serving models at the top at the Model Versions tab of the Serving page of the registered model in the MLflow UI.

You can test the deployed model either by making a REST API request to one of the available endpoints or using the **Model Serving** UI in the Azure workspace.

Azure Databricks creates unique clusters when a registered model is enabled to be served and deploys all the versions of the model except the ones archived. If any error occurs, Azure Databricks will restart the cluster, and if you disable the serving it will terminate it. As mentioned before, this model serving also updates automatically when we update the version of a model that is being served. To be able to query these endpoints, we just need to authenticate ourselves using a generated Databricks access token.

The MLflow model serving is indented for low throughput and is non-critical, with a limitation of 16 **megabytes** (**MB**) per request made. For applications that require higher throughput, you should consider using batch predictions. The clusters used are all-purpose clusters, so they will have the same configurations of scalability as these ones.

Consider that a cluster is maintained even if no active versions of the model exist, so you should always be sure to disable the serving of a registered model if it's no longer needed.

Scoring a model

The most straightforward method to score an MLflow deployed model is to test the model in the Serving UI. There, you can insert data structured as **JavaScript Object Notation (JSON)** format and use it to test the model by clicking the Send Request button. This way, we can ensure that the model returns a valid prediction.

The next step is to obviously consume the predictions made by this model by making REST API requests to the available endpoints.

For example, we can make a request to a model using a `curl` bash command. For example, for a given model we can structure the request, as shown in the following code snippet:

```
curl -X POST -u token:$DATABRICKS_API_TOKEN $MODEL_VERSION_URI \
  -H 'Content-Type: application/json' \
  -d '[
    {
      "sepal_length": 4.2,
      "sepal_width": 3.1,
      "petal_length": 2.1,
      "petal_width": 0.1
    }
  ]'
```

In this example, we are passing a request to a simple classifier trained on the Iris dataset. You should replace the `DATABRICKS_API_TOKEN` and `MODEL_VERSION_URI` placeholders in the endpoint with your generated Databricks API token and Model URI respectively.

Here, we are passing JSON as key-value structured data and receiving the response from the model endpoint.

We can also do this programmatically using Python. In the next example, we will make a request to the endpoint using Pandas DataFrames and NumPy arrays:

1. First, we will define a function to serialize the data into JSON, as follows:

    ```
    import numpy as np
    import pandas as pd
    def create_tf_serving_json(data):
    ```

```
    return {'inputs': {name: data[name].tolist() for name
in data.keys()} if isinstance(data, dict) else data.
tolist()}
```

2. Then, we can create a Pandas data frame with the data that we will pass to the model to make predictions, like this:

```
data =  pd.DataFrame([{
        "sepal_length": 4.2,
        "sepal_width": 3.1,
        "petal_length": 2.1,
        "petal_width": 0.1
}])
```

3. Now, we can serialize this data as JSON and send it as a POST request to the model endpoint, as follows:

```
import requests
headers = {
    "Authorization": f"Bearer {databricks_token}",
    "Content-Type": "application/json",
    }
data_json = data.to_dict(orient='records')
response = requests.request(method='POST',
                            headers=headers,
                            url=model_uri,
                            json=data_json)
if response.status_code != 200:
raise Exception(f"Request failed with status
{response.status_code}, {response.text}")
print(response.json())
```

This way, we can score a model using Pandas Data Frames, but we can also use NumPy arrays as both have a method to serialize the data into dictionaries that can easily be converted into JSON format.

4. Here, we will pass the data as a NumPy array:

```
data = np.array([[5.1, 3.5, 1.4, 0.2]])
data_json = create_tf_serving_json(data)
response = requests.request(method='POST',
                            headers=headers,
                            url=model_uri,
                            json=data_json)
if response.status_code != 200:
raise Exception(f"Request failed with status
{response.status_code}, {response.text}")
print(response.json())
```

In the **Serving** tab of the Model Registry UI, you will be prompted with a status of the serving clusters and data about the individual model versions available, along with their URIs. You can check the status of a cluster in the **Model Events** tab, which also holds data about all events associated with that model. You can also query specific versions events in the **Model Versions** tab.

Summary

The development process of machine learning models can be a complicated task because of the inherent mixed background of the discipline and the fact that it is commonly detached from the common software development lifecycle. Moreover, we will encounter issues when transitioning the models from development to production if we are not able to export the used preprocessing pipeline that was used to extract features of the data.

As we have seen in this chapter, we can tackle issues using MLflow to manage the model lifecycle and apply staging and version control to the models used, and effectively serialize the preprocessing pipeline to be used to preprocess data to be inferred.

In the next chapter, we will explore the concept of distributed learning, a technique in which we can distribute the training process of deep learning models to many workers effectively in Azure Databricks.

12
Distributed Deep Learning in Azure Databricks

In the previous chapter, we have learned how we can effectively serialize machine learning pipelines and manage the full development life cycle of machine learning models in Azure Databricks. This chapter will focus on how we can apply distributed training in Azure Databricks.

Distributed training of deep learning models is a technique in which the training process is distributed across workers in clusters of computers. This process is not trivial and its implementation requires us to fine-tune the way in which the workers communicate and transmit data between them, otherwise distributing training can take longer than single-machine training. Azure Databricks Runtime for Machine Learning includes Horovod, a library that allows us to solve most of the issues that arise from distributed training of deep learning algorithms. We will also show how we can leverage the native Spark support of the TensorFlow machine learning framework to train deep learning models in Azure Databricks.

The concepts that we will go through in the chapter are listed as follows:

- Distributed training for deep learning
- Using the Horovod distributed learning library in Azure Databricks
- Using the Spark TensorFlow Distributor package

Before we dive into these concepts, let's go through the requirements in order to execute the examples shown here.

Technical requirements

In this chapter, we will use concepts and apply tools related to common topics in data science and optimization tasks. We will assume that you already have a good understanding of concepts such as neural networks and hyperparameter tuning, and also general knowledge of machine learning frameworks such as TensorFlow, PyTorch, or Keras.

In this chapter, we will discuss various techniques in order to distribute and optimize the training of deep learning models using Azure Databricks, so if you are not familiar with these terms, it would be advisable for you to review the TensorFlow documentation on how neural networks are designed and trained.

In order to work on the examples given in this chapter, you will need to have the following:

- An Azure Databricks subscription.
- An Azure Databricks notebook attached to a running cluster with Databricks Runtime ML version 7.0 or higher.

Distributed training for deep learning

Deep neural networks (**DNNs**) have driven the advancement of **artificial intelligence** (**AI**) in the last decades in areas such as computer vision and neural network processing. These are applied every day to solve challenges in diverse use cases.

In order to scale the performance of models, it is necessary to develop complex model architectures with millions of trainable parameters, making the computations required for the training a resourceful operation. As the amount of available data to train models increases, we need to scale up the training pipeline of deep learning models in order to be able to use this available data..

Commonly, in order to train a DNN, we need to follow three basic steps, which are listed here:

1. Pass the data through the layers of the network to compute the model loss in an operation called forward propagation.

2. Backpropagate this loss from the output layer to the first layer in order to compute the gradients.

3. We use calculated gradients in order to update the model weights in the optimization step of the training process.

The scalability of the training process of a DNN can be achieved in the following two ways:

- Using data parallelization
- Using model parallelization

A data parallelization strategy involves copying the model into each one of the available workers and fitting them, using assigned portions of the data. Later on, the workers communicate with each other in order to properly update the model's weights, using the average calculated gradients.

Although it might sound simple, the implementation can be quite complex, and the complexities arising from tackling the operation in this way can completely erase the performance gain from distributing the training, if it's not done properly.

Modern neural networks have grown at a fast pace in the last decade. Some of the architectures used in image processing have millions of parameters that need to be trained on massive amounts of data. This computationally expensive training process has led to the development of techniques to parallelize the process into several workers.

As the amount of data required to train the model increases, we can increase the number of workers to run the computations. Commonly, workers receive a copy of the neural network, and mini-batches are distributed to each one of them to calculate a gradient, based on the training batch of data. These gradients are averaged to recalculate the new weights of the neural network.

When the data increases and the models have millions of trainable parameters, these gradients can take **gigabytes (GB)** of space in the memory, and the communication between workers starts to become a problem as we reach the maximum bandwidth of the network.

The other impediment that we face is that the way in which the gradients are commonly calculated involves waiting until all the gradients from every worker have been calculated. One alternative to solve this problem is to introduce algorithms that run asynchronous calculations of gradients, but these tend to be very hard to debug.

This is when a technique derived from high-performance computing, the *ring allreduce* technique, comes into play. This allows us to improve the performance of the training process of neural networks by making use of the optimal bandwidth ring allreduce.

The ring allreduce technique

As we introduce new workers in order to handle increases in the size of the calculated gradients, we start to notice that the communication costs start to grow linearly. We can configure the communication between the workers in order to find the optimal strategy to make use of the available bandwidth. To do this, we use the ring all reduce, a communication algorithm where the communication cost is held constant no matter the number of workers introduced, and it's only dependent on the slowest connection in the system.

The `ring allreduce` algorithm orders the workers in a logical ring, where each worker can only send data to its right neighbor and receive data from the one at its left. The algorithm is depicted in the following diagram:

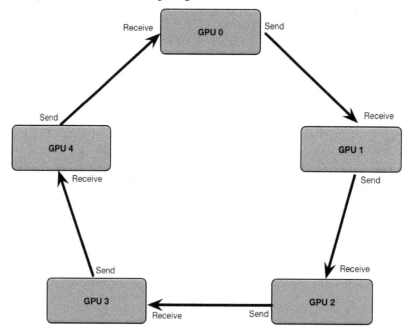

Figure 12.1 – How the Ring All Reduce algorithm allows communication between workers

The `ring allreduce` algorithm works in two steps. First, each worker exchanges data so that each worker has a chink of the result, and then runs an `allgather` operation so that all the workers have a complete result.

Another point where we can improve the communication between workers is by making use of the inherent calculation process of the gradients in a neural network.

To train a neural network, we first run a forward propagation and then a backward propagation to compute the gradients at each of the layers of the neural network, starting from the output layer up to the input layer. Therefore, the first available gradients are the ones from the output layer. We can exploit this process by sending the gradients as soon as they are ready instead of waiting for all of them to be finished, in order to avoid spikes in bandwidth use and overlays in communication with the rest of the workers. This way, we overcome the inconvenience of having to wait for each worker to finish the gradient computation over all of the layers of the neural network.

The good thing about this approach is that it is agnostic from the network architecture and framework used to create the model, and it allows us to better harness the power of parallel computing for distributed training of deep learning models.

The `horovod` library applies the `ring allreduce` algorithm to enable distributed training of deep learning networks in distributed computing systems such as Azure Databricks. In the next sections, we will learn how we can leverage the use of `HorovodRunner` to efficiently train deep learning models.

Using the Horovod distributed learning library in Azure Databricks

`horovod` is a library for distributed deep learning training. It supports commonly used frameworks such as TensorFlow, Keras, and PyTorch. As mentioned before, it is based on the `tensorflow-allreduce` library and implements the `ring allreduce` algorithm in order to ease the migration from single-**graphics processing unit (GPU)** training to parallel-GPU distributed training.

In order to do this, we adapt a single-GPU training script of a deep learning model to use the `horovod` library during the training process. Once we have adapted the script, it can run on single or multiple GPUs without changes to the code.

The `horovod` library uses a data parallelization strategy by allowing efficient distribution of the training to multiple GPUs in parallel in an optimized way, by implementing the `ring allreduce` algorithm to overcome communication limitations.

It is implemented in a way that each GPU gets a mini-batch of data in order to independently calculate a gradient. For example, if we have a batch size of 32 and two available GPUs, each one will receive a set of 16 records to calculate the gradient update. These gradient updates are later on synchronized and averaged between all the available GPUs in order to finally use this to update the model weights.

In each one of the training runs over the batches of data, `horovod` follows the process outlined next in order to handle the distributed computation among GPUs:

- Each worker receives a copy of the model and the dataset.

- Once the go signal is received, the worker computes the gradients on an assigned batch of the data.

- The `ring allreduce` algorithm is used to synchronize and average the calculated gradients across GPUs.

- All the workers update their models' weights using the calculated averaged gradient.

One of the most crucial aspects of this operation is to maintain the same model across workers. If the models diverged somehow, this would render inconsistent gradient calculations. This is where the `ring allreduce` algorithm plays an important role in synchronizing the tensors representing the model-trainable parameters in an efficient way.

In order to develop a training pipeline that can be efficiently used for distributed training of deep learning models, a common workflow is outlined next:

- Create a training script for a single machine using one of the Horovod supported frameworks such as TensorFlow, Keras, or PyTorch.

- Add the `hvd.init()` function to initialize the `horovod` library.

- Choose one of the available GPU servers using the `config.gpu_options.visible_device_list` option.

- Use `hvd.DistributedOptimizer` to wrap the optimizer to distribute the gradient descent calculation across workers.

- Use `hvd.BroadcastGlobalVariablesHook(0)` to distribute initial variables' state to ensure consistency in the initialization of all the random weights of the workers' models. Another alternative to do this is to use `hvd.broadcast_global_variables()` to broadcast global starting variables.

- Set your code to save the checkpoints in the 0 worker to avoid any corruption of data due to overlapping read/write operations.

- Use `HorovodRunner` to wrap a single Python function that packs the training procedure.

`horovod` is a library available in Azure Databricks Runtime for machine learning. In the next sections of the chapter, we will learn how we can use it to distribute the training of deep learning models and to effectively parallelize the tuning of hyperparameters.

As an example, we will show a construct of how we must modify our code to be able to run in distributed mode. We will use a toy model example with a TensorFlow model. This assumes that we have a TensorFlow v1 model that we want to train as a distributed model. The steps are:

1. To do this, we would need to first make the imports of TensorFlow and Horovod and initialize Horovod, as follows:

   ```
   import tensorflow as tf
   import horovod.tensorflow as hvd
   hvd.init()
   ```

2. Next, we will ping the GPU that will be used to process the local rank of the node, like this:

   ```
   config = tf.ConfigProto()
   config.gpu_options.visible_device_list = str(hvd.local_
   rank())
   ```

3. After this point, we can start to build the model and define the `loss` function that will be utilized. We will abstract this point and assume that you already have a `loss` function to optimize.

4. The next step is to apply the Horovod distributed optimizer to the selected optimizer in order to distribute the training of the gradient calculation. As an example, we will define a TensorFlow `AdagradOptimizer` optimizer and set the learning rate to be a variable dependent on the size that Horovod passes to the workers, a parameter used to control the learning rate. The code for this is shown in the following snippet:

   ```
   opt = tf.train.AdagradOptimizer(0.01 * hvd.size())
   ```

5. Now, we can use the Horovod distributed optimizer in order to wrap the selected optimizer, as follows:

   ```
   opt = hvd.DistributedOptimizer(opt)
   ```

6. It is important to add a hook to broadcast the initiation variables from rank 0 to all other processes during initialization to be able to replicate structures that depend on random initialization parameters, such as initialization of model weights. The code to do this is illustrated in the following snippet:

    ```
    hooks = [hvd.BroadcastGlobalVariablesHook(0)]
    ```

7. At this point, we are ready to start the training operation using our predefined loss function, as follows:

    ```
    train = opt.minimize(loss)
    ```

8. We need to specify the checkpoints to be saved only on worker 0 to prevent other workers from overlapping write operations between them. We run the following code to do so:

    ```
    checkpoint_dir = '/tmp/training_logs' if hvd.rank() == 0
    else None
    ```

9. Now, we will use the `MonitoredTrainingSession` checkpoint to take care of the operations required for session initialization, loading, and saving to and from checkpoints, and closing when the operation has been completed or if any error occurs. The code for this is shown in the following snippet:

    ```
    with tf.train.\
    MonitoredTrainingSession(checkpoint_dir=checkpoint_dir,
                            config=config,
                            hooks=hooks) as mon_sess:
      while not mon_sess.should_stop():
        mon_sess.run(tr)
    ```

This way, we can modify a TensorFlow v1 training script to use Horovod for distributed training. Here, Horovod will dynamically modify the batch size and the learning rate in each one of the workers and will process the training, synchronously scaling with the number of available workers. The `hvd.DistributedOptimizer` optimizer will take care of calculating the gradients on batches of data in each one of the workers and modifying the model parameters based on the averaged gradients.

Finally, we have made sure to broadcast all of the initiation variables to ensure consistency in the models across workers, and enforced the checkpoints to be written by the rank 0 workers to prevent any corruption of the checkpoint.

In the next sections, we will dive into more concrete examples of how we can use Horovod for distributed training of deep learning models in **central processing unit** (**CPU**) or GPU clusters.

Installing the horovod library

The horovod library comes pre-installed in Azure Databricks Runtime for Machine Learning, so in most cases you won't need to install it unless you are using a common Azure Databricks Runtime. Nevertheless, one thing to mention is that if you upgrade or downgrade one of the frameworks used to build the model—such as TensorFlow, Keras, or PyTorch—you will need to reinstall the horovod library and compile it again.

In cases where we need to upgrade or downgrade one of the frameworks, we might encounter import errors, indicating that horovod was installed before the required framework, such as TensorFlow, Keras, or PyTorch. This is due to the fact that Horovod is compiled during the installation, and if the required library was not present during the installation, it is not compiled.

If we have reinstalled one of the required libraries, we can overcome this problem by following the next steps:

1. Check that the required library was correctly installed.
2. Uninstall horovod using Python pip.
3. Install the CMake Python library with pip.
4. Reinstall horovod using Python pip.

This way, we can ensure that the horovod library was compiled once the required framework was installed.

Once we have the horovod library available to use in our environment, we will center our attention on which are the available features that we can leverage to efficiently use it for distributed training in Azure Databricks.

Using the horovod library

In this section, we will show how we can apply horovod in order to train a simple classifier on the **Modified National Institute of Standards and Technology** (**MNIST**) dataset using the Keras TensorFlow **application programming interface** (**API**) and then adapt the code in order for this to be trained in a distributed way. First, we will create two functions that will be used to create the dataset and to build the model.

In order to track runs and processes, the `horovod` library uses process IDs and the number of processes as parameters. These parameters are the rank and the size, which will be used later on by `HorovodRunner` to track and control the process.

Training a model on a single node

We will create a function that will take as arguments the rank and size and return the data ready to be used to train the model. This function works both on a single node or during distributed learning. The steps are as follows:

1. We will use the `keras.datasets.mnist.load_data()` function to get the dataset and then resize and preprocess it for it to be able to be directly used to train the model. Note in the following code snippet that the imports are inside the function:

```
def get_dataset(num_classes, rank=0, size=1):
    from tensorflow import keras
    (x_train, y_train), (x_test, y_test) = keras.datasets.
mnist.load_data('MNIST-data-%d' % rank)
    x_train, y_train = \
    x_train[rank:size],y_train[rank:size]
    x_test,y_test = x_test[rank:size],y_test[rank:size]
    x_train = x_train.reshape(x_train.shape[0], 28, 28, 1)
    x_test = x_test.reshape(x_test.shape[0], 28, 28, 1)
    x_train = x_train.astype('float32')/255
    x_test = x_test.astype('float32')/255
    y_train = keras.utils.to_categorical(y_train,
                               num_classes)
    y_test = keras.utils.to_categorical(y_test,
                               num_classes)
    return (x_train, y_train), (x_test, y_test)
```

This function allows us to load the dataset in nodes and preprocess it into the required format to be passed to the model for training.

2. The next step is to define a function to create a model that will be trained.
    We will define a Keras **convolutional neural network (CNN)** with dropout after
    the flattening of the max pooling and after the final softmax layer. We will use
    a hidden dense layer of 64 neurons. Note in the following code snippet that we have
    the imports inside the functions to import the modules to the workers later:

```
def get_model(num_classes):
   from tensorflow.keras import models
   from tensorflow.keras import layers
   model = models.Sequential()
   model.add(layers.Conv2D(32, 3,
                   activation='relu',
                   input_shape=(28, 28, 1)))
   model.add(layers.MaxPooling2D())
   model.add(layers.Dropout(0.1))
   model.add(layers.Flatten())
   model.add(layers.Dense(64, activation='relu'))
   model.add(layers.Dropout(0.2))
   model.add(layers.Dense(num_classes,
                   activation='softmax'))
   return model
```

3. Now, we can run the training on a single node using the functions that we have just
    defined. We will create a train function that will be able to be executed both on
    a single node and a distributed set of workers. Consider that the learning rate must
    be a parameter in the function for Horovod to control the learning rate in each one
    of the workers during the distributed training process. The code is shown in the
    following snippet:

```
def train(learning_rate=0.1,
batch_size = 64,
epochs = 10,
num_classes = 10):
   from tensorflow import keras
   (x_train, y_train), (x_test, y_test) = \
   get_dataset(num_classes)
   model = get_model(num_classes)
```

```
optimizer = keras.optimizers.Adadelta(lr=learning_rate)
model.compile(optimizer=optimizer,
              loss='categorical_crossentropy',
              metrics=['accuracy'])
model.fit(x_train, y_train,
          batch_size=batch_size,
          epochs=epochs,
          verbose=2,
          validation_data=(x_test, y_test))
return model
```

4. Now, we can run the training on a single node, using the functions that we have just defined. In order to get a trained model, we just need to call the function specifying the required parameters, as shown in the following code snippet:

```
model = train(learning_rate=0.1,
batch_size = 64,
epochs = 10,
num_classes = 10)
```

5. The process will take a couple of minutes to train, and afterward the model will be returned, ready to be used. Once the process is finished, we can test it by calculating the model accuracy on the test data by using the previously defined get_dataset() function, as follows:

```
_, (x_test, y_test) = get_dataset(num_classes)
loss, accuracy = model.evaluate(x_test, y_test,
                                batch_size=128)
print("loss:", loss)
print("accuracy:", accuracy)
```

We can now see the loss and accuracy of the model in the test set, as illustrated in the following screenshot:

Cmd 5

```
1  num_classes = 10
2  _, (x_test, y_test) = get_dataset(num_classes)
3  loss, accuracy = model.evaluate(x_test, y_test, batch_size=128)
4  print("loss:", loss)
5  print("accuracy:", accuracy)
6

79/79 [==============================] - 1s 8ms/step - loss: 0.0808 - accuracy: 0.9752
loss: 0.08081278949975967
accuracy: 0.9751999974250793
```

Figure 12.2 – The loss and accuracy of the model in the test set

The way in which the get_dataset() and get_model() functions are designed made it easy not only to retrain on new parameters but also to be reusable when these are distributed to the nodes, which is what we will do in the next section.

Distributing training with HorovodRunner

In order to distribute the training to the workers, we will start by creating a checkpoint location for HorovodRunner in the **Databricks File System (DBFS)** ml-fuse directory available in Azure Databricks Runtime for Machine Learning, as follows:

```
import os
import time
checkpoint_dir = '/dbfs/ml/MNISTDemo/train/{ time.time()}/'
os.makedirs(checkpoint_dir)
print(checkpoint_dir)
```

This directory is available to all the workers and it's optimized for machine learning I/O operations on training data. The directory can be seen in the following screenshot:

Cmd 6

```
1  import os
2  import time
3  checkpoint_dir = f'/dbfs/ml/MNISTDemo/train/{ time.time()}/'
4  os.makedirs(checkpoint_dir)
5  print(checkpoint_dir)
6

/dbfs/ml/MNISTDemo/train/1617111715.7711205/
```

Figure 12.3 – The ml directory

The next step is to define a function that will adapt the training function we have previously defined to be able to be distributed to the workers. The function itself packs a lot of operations, therefore we can explain the steps that it follows as the following:

1. It imports the required modules into each of the workers.

2. It initializes horovod using the horovod.tensorflow.keras.init() function.

3. It pings the GPU to be used if the run is local and if there is a GPU available; otherwise, it will skip this step.

4. It uses the get_dataset() and get_model() functions that were defined earlier for horovod to dynamically adjust the learning rate based on the available GPUs, rank, and size.

5. It applies the Horovod distributed optimizer in order to distribute and compute the gradients during the distributed training.

6. It creates a callback process to distribute the state variables from the rank 0 to all the other processes in order to ensure consistency of the worker model states when the process has randomly initiated weights.

7. It enforces to save the checkpoints in the 0 worker to prevent conflicts or corruption of data due to overlapping read/write operations.

We will define the Horovod training function for distributed training, as follows:

```
def train_hvd(checkpoint_path, learning_rate=1.0):
from tensorflow.keras import backend as K
from tensorflow.keras.models import Sequential
import tensorflow as tf
from tensorflow import keras
    import horovod.tensorflow.keras as hvd
    hvd.init()
    gpus = tf.config.experimental.list_physical_devices('GPU')
    for gpu in gpus:
        tf.config.experimental.set_memory_growth(gpu, True)
    if gpus:
        tf.config.experimental.set_visible_devices(gpus[hvd.
local_rank()], 'GPU')


    (x_train, y_train), (x_test, y_test) = get_dataset(num_
```

```
classes, hvd.rank(), hvd.size())
    model = get_model(num_classes)
        optimizer = keras.optimizers.Adadelta(lr=learning_rate *
hvd.size())
        optimizer = hvd.DistributedOptimizer(optimizer)
    model.compile(optimizer=optimizer,
                    loss='categorical_crossentropy',
                    metrics=['accuracy'])
    callbacks = [
        hvd.callbacks.BroadcastGlobalVariablesCallback(0),
    ]
        if hvd.rank() == 0:
            callbacks.append(keras.callbacks.ModelCheckpoint(
checkpoint_path,
save_weights_only = True))
    model.fit(x_train, y_train,
                batch_size=batch_size,
                callbacks=callbacks,
                epochs=epochs,
                verbose=2,
                validation_data=(x_test, y_test))
```

We will use this function in order to distribute the training to the workers. We will set an np parameter in HorovodRunner that relates to the specified number of parameters—a value that is dependent on the number of available workers. So, if we specify np=1, HorovodRunner will train on the driver node on a single-process execution.

In this example, we will set the np value to 2:

```
from sparkdl import HorovodRunner
checkpoint_path = checkpoint_dir + '/checkpoint-{epoch}.ckpt'
hr = HorovodRunner(np=2)
hr.run(train_hvd, checkpoint_path=checkpoint_path,
        learning_rate=0.1)
```

The process will be executed by `HorovodRunner` by transforming the defined functions to get the dataset and create the model, and then transform them into pickle files that will then be distributed to the workers, as illustrated in the following screenshot:

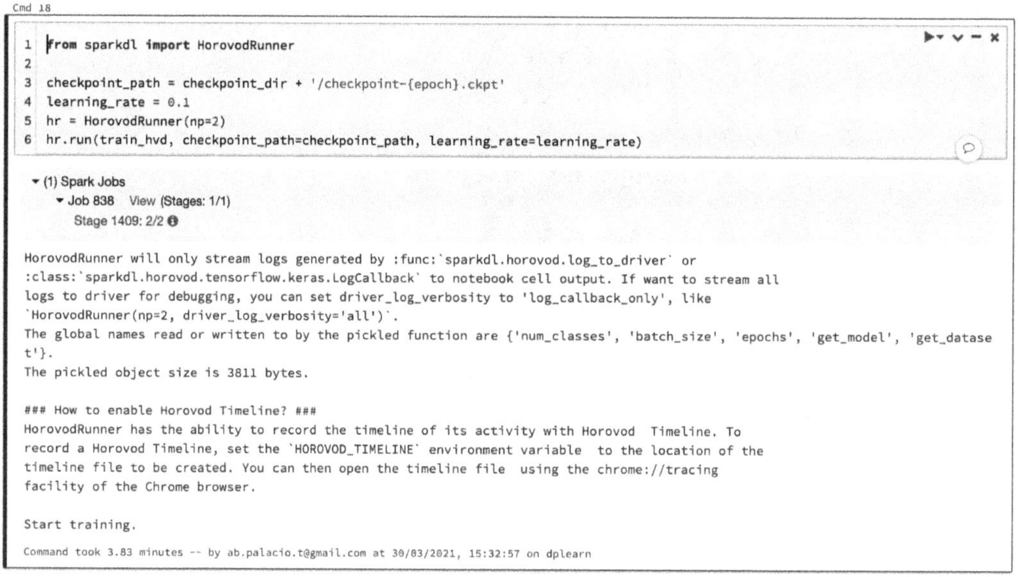

```
Cmd 18
1  from sparkdl import HorovodRunner
2
3  checkpoint_path = checkpoint_dir + '/checkpoint-{epoch}.ckpt'
4  learning_rate = 0.1
5  hr = HorovodRunner(np=2)
6  hr.run(train_hvd, checkpoint_path=checkpoint_path, learning_rate=learning_rate)

▼ (1) Spark Jobs
    ▼ Job 838    View (Stages: 1/1)
        Stage 1409: 2/2 ⓘ

HorovodRunner will only stream logs generated by :func:`sparkdl.horovod.log_to_driver` or
:class:`sparkdl.horovod.tensorflow.keras.LogCallback` to notebook cell output. If want to stream all
logs to driver for debugging, you can set driver_log_verbosity to 'log_callback_only', like
`HorovodRunner(np=2, driver_log_verbosity='all')`.
The global names read or written to by the pickled function are {'num_classes', 'batch_size', 'epochs', 'get_model', 'get_datase
t'}.
The pickled object size is 3811 bytes.

### How to enable Horovod Timeline? ###
HorovodRunner has the ability to record the timeline of its activity with Horovod  Timeline. To
record a Horovod Timeline, set the `HOROVOD_TIMELINE` environment variable  to the location of the
timeline file to be created. You can then open the timeline file  using the chrome://tracing
facility of the Chrome browser.

Start training.

Command took 3.83 minutes -- by ab.palacio.t@gmail.com at 30/03/2021, 15:32:57 on dplearn
```

Figure 12.4 – Dataset transformed to pickle files

The workers in turn read and deserialize the functions and run them in order to train the model.

After the process has been completed, we can load the model trained by `HorovodRunner` and use it to run predictions on new data. We can use the Keras model that we have previously defined using the `tf.train.latest_checkpoint()` function to load the latest saved checkpoint file, as follows:

```
import tensorflow as tf
hvd_model = get_model(num_classes)
hvd_model.compile(optimizer=tf.keras.optimizers.
Adadelta(lr=learning_rate),
                  loss='categorical_crossentropy',
                  metrics=['accuracy'])
hvd_model.load_weights(tf.train.latest_checkpoint(os.path.
dirname(checkpoint_path)))
```

We can test that the model has been correctly loaded by running predictions on the test dataset, as follows:

```
num_classes = 10
_, (x_test, y_test) = get_dataset(num_classes)
loss, accuracy = hvd_model.evaluate(x_test, y_test,
                                    batch_size=128)
print("loaded model loss and accuracy:", loss, accuracy)
```

In this way, and by making small but meaningful changes to the code, we have adopted a single machine-training script in order to be able to use it for distributed learning in Azure Databricks, running on both CPU and GPU clusters.

In the next section, we will learn how to leverage the advantages of the `horovod` library in order to tune the model's hyperparameter.

Distributing hyperparameter tuning using Horovod and Hyperopt

As we have seen previously, we can improve the performance of the model by tuning the hyperparameter using Hyperopt. Hyperopt evaluated the best hyperparameter by generating trials for different points in the defined search space, using an adaptive strategy. The evaluations take advantage of the distributed nature of Azure Databricks.

The Hyperopt trials are evaluated on a single node while `HorovodRunner` is launched from the driver node, and distributes the training to the workers. We will learn how we can use `HorovodRunner` in order to increase the performance of the model tuning, using distributed learning to run the trials.

The first objective is to define a function that will be passed to `HorovodRunner` in order to run the distributed learning. To do this, we will define a function to minimize, specify a search space of hyperparameters, and use the `Hyperopt fmin()` function to tune the model. The change compared to the example in the previous section is that this training will be optimized to run a distribution, using `HorovodRunner`. The next train function is passed to the `Hyperopt fmin()` function, to be the target function to be minimized. It is configured to return a dictionary with key values named `loss` and `Status`, as required by the `Hyperopt fmin()` function.

Here, we are using the previously defined `train_hvd()` function:

```
from hyperopt import fmin, tpe, hp, Trials, STATUS_OK
from sparkdl import HorovodRunner


def wrap_train(params):
  hr = HorovodRunner(np=2)
  loss, acc = hr.run(train_hvd,
                     learning_rate=params['learning_rate'],
                     batch_size=params['batch_size'],
                     checkpoint_dir=checkpoint_dir)
  return {'loss': loss, 'status': STATUS_OK}
```

The next step is to define the search space over which we will scan for the best set of hyperparameters. Here, we will use the Hyperopt optimization algorithm to tune the learning rate and batch size:

```
import numpy as np
space = {
  'learning_rate': hp.loguniform('learning_rate',
                                  np.log(1e-4),
                                  np.log(1e-1)),
  'batch_size': hp.choice('batch_size', [32, 64, 128]),
}
```

Now that we have defined the objective function and the search space that will be scanned, we can use the `Hyperopt fmin()` function to run the hyperparameter optimization. While Hyperopt will take advantage of the parallel execution on each of the workers, the `horovod` library will distribute the process to each of the available workers.

To run the optimization, we must use the `max_eval` parameter in order to specify the number of points that will be evaluated for each defined test case in the search space, generating a model as a trial, and use it to evaluate the model in that point of the search space.

In this case, we will be using as a search algorithm `hyperopt.tpe.suggest`, which is the Tree of Parzen Estimator, an adaptive approach that selects the hyperparameter to be tested based on previous results. Given this, we can find the best parameters by running the following code block:

```
best_param = fmin(
    fn=wrap_train,
    space=space,
    algo=tpe.suggest,
    max_evals=8,
    return_argmin=False,
)
```

Once the target function to be minimized has been optimized, we can print out the parameters that produced the best model, as follows:

```
print(best_param)
```

One important thing to notice is that in the `Hyperopt fmin()` function, we are not using a trials parameter. This is because when Hyperopt is used in conjunction with `HorovodRunner`, we don't need to pass a trials argument because Hyperopt will use the default `Trials` class instead of `SparkTrials`. This is because the `SparkTrials` class is not usable in distributed training because the evaluations are done on a single node.

In this simple example, we have extended the functionality of `HorovodRunner` in order to use it for hyperparameter tuning of the CNN TensorFlow model that was originally trained on a single node. To do this, we have created a target objective function to be minimized by the `hyperopt` library, defined a search space based on the hyperparameters that we want to scan, and run hyperparameter tuning using a cluster with two workers running Azure Databricks Runtime for Machine Learning.

The next section will introduce Spark TensorFlow Distributor, a library that allows us to distribute train TensorFlow machine learning models in Azure Databricks.

Using the Spark TensorFlow Distributor package

One of the most commonly used frameworks in deep learning is the TensorFlow library, which also supports distributed training on both CPU and GPU clusters. We can use it to train deep learning models in Azure Databricks by using Spark TensorFlow Distributor, which is a library that aims to ease the process of training TensorFlow models with complex architecture and lots of trainable parameters in distributed computing systems with large amounts of data.

Spark was limited to distributed training because of the standard execution mode, which is the **Map/Reduce** mode. In this mode, jobs are executed independently in each worker without any communication between them. In Spark 3.0, there is a new execution mode named barrier execution that allows us to easily train deep learning models in a distributed way by allowing communication between workers during the execution.

Spark TensorFlow Distributor is a TensorFlow-native package that makes use of the barrier execution mode, allowing workers to share data between each other during the execution. It is already installed on Azure Databricks Runtime for Machine Learning, but you can also do the install using `pip` on a cluster, like this:

```
pip install spark-tensorflow-distributor
```

Before the implementation of the barrier execution mode, it was necessary to manage **Map/Reduce-based** jobs, which made the implementation very inefficient for distributed training of deep learning models.

The Spark TensorFlow Distributor library is built around the `tensorflow. distribute.Strategy` TensorFlow API, which is designed to distribute training in GPUs or CPU clusters. It can be used to distribute the training while making minor changes to the training script. It was built in order to do the following:

- Ease the transition from single machine training to distributed training on a cluster.
- Distribute custom training scripts made using TensorFlow.

In this section, we will demonstrate how we can implement a distributed training strategy for TensorFlow deep learning models using the `MirroredStrategyRunner` class of the Spark TensorFlow Distributor library.

To use the Spark TensorFlow Distributor library, we just need to provide a `train()` function with a certain structure, and the package will handle the necessary configurations.

We will start by first making an import of the TensorFlow library and setting the initial parameters, which are the number of workers and available GPUs. We will assume that the driver and nodes have the same types of instances, therefore the calculation of total available GPUs will be the number of GPUs in a worker times the number of instances in the cluster. The code is shown in the following snippet:

```
import tensorflow as tf


NUM_WORKERS = 2
CLUSTER_GPUS = len(tf.config.list_logical_devices('GPU')) *
NUM_WORKERS
USE_GPU = CLUSTER_GPUS > 0
```

Next, we will create a training function that will run on a single machine before we distribute the training to the workers. It is important to note that it is recommended to make the imports inside the function. This is because the training function will be serialized to a pickle and this way, we avoid any possible issue in this process. In the function, the workflows are implemented next, as follows:

1. We define and call a function to load the MNIST dataset and transform it to a normalized dataset of tensors.

2. We define and call a function to create and compile the model that will be trained.

3. We define the `autoshard` data policy, which is a TensorFlow parameter used in distributed learning to specify the way in which the way. If the option is set to `FILES`, the data will be distributed into the workers partitioned by files. If the option is set to `DATA`, all workers receive the full dataset but will only work on the portion assigned to them.

4. We fit the model on the train data.

Let's define the training function using a default learning rate, buffer, and batch size, as follows:

```
def train(BUFFER_SIZE = 5000, BATCH_SIZE = 32, learning_
rate=0.005):
    import tensorflow as tf
    import uuid


    def get_datasets():
        (mnist_images, mnist_labels), _ = \
            tf.keras.datasets.mnist.load_data(path=str(uuid.
```

```
uuid4())+'mnist.npz')

    dataset = tf.data.Dataset.from_tensor_slices((
        tf.cast(mnist_images[..., tf.newaxis] / 255.0,
            tf.float32),
        tf.cast(mnist_labels, tf.int64))
    )
    dataset = dataset.repeat().shuffle(BUFFER_SIZE).
batch(BATCH_SIZE)
    return dataset
```

Now that we have defined the function that will load, preprocess, and return the training data, we can define a function that will create the Keras CNN, as follows:

```
def get_model():
    model = tf.keras.Sequential([
        tf.keras.layers.Conv2D(32, 3, activation='relu',
                                input_shape=(28, 28, 1)),
        tf.keras.layers.MaxPooling2D(),
        tf.keras.layers.Flatten(),
        tf.keras.layers.Dense(128, activation='relu'),
        tf.keras.layers.Dense(10, activation='softmax'),
    ])
    model.compile(
        loss=tf.keras.losses.sparse_categorical_crossentropy,
        optimizer=tf.keras.optimizers.SGD(learning_rate=learning_
rate),
        metrics=['accuracy'],
    )
    return model
train_datasets = get_datasets()
worker_model = get_model()
options = tf.data.Options()
options.experimental_distribute.auto_shard_policy = \
    tf.data.experimental.AutoShardPolicy.DATA
train_datasets = train_datasets.with_options(options)
worker_model.fit(x=train_datasets, epochs=3,
                steps_per_epoch=5)
```

Once we have defined the function that will handle the model creation, compilation, and training, we can run the training using the Spark TensorFlow Distributor `MirroredStrategyRunner` class in local mode. This mode will train the model in all the driver-available GPUs. The code is shown in the following snippet:

```
from spark_tensorflow_distributor import MirroredStrategyRunner
runner = MirroredStrategyRunner(num_slots=1,
                                local_mode=True,
                                use_gpu=USE_GPU)
runner.run(train)
```

In this very simple way, we can execute the code in the local node by setting the `local_mode` parameter to `True`. Now, in order to run the distributed training, we will set the `local_mode` parameter to `False` as we want our function to run on the workers of the Spark cluster.

We will define a number of slots that depend on whether we have a GPU available or not. If we have a GPU available, the number of slots will be set to the number of available GPUs. If we are using a CPU cluster, we should set the number of slots to a reasonable number, which in our case is 4, as can be seen in the following code snippet:

```
NUM_SLOTS = CLUSTER_GPUS if USE_GPU else 4
runner = MirroredStrategyRunner(num_slots=CLUSTER_GPUS,
                                use_gpu=USE_GPU)
runner.run(train)
```

The `MirroredStrategyRunner` class is a strategy designed for nodes that have multiple GPUs available, but we can also specify a custom strategy for the distributed training. To do so, you need to create a `tf.distribute.Strategy` object within the `train()` function and set `use_custom_strategy` to `True` in the `MirroredStrategyRunner` function.

As an example, we can change the distributed training strategy to make use of `tf.distribute.experimental.MultiWorkerMirroredStrategy`. The `MultiWorkerMirroredStrategy` strategy is designed to distribute the training across workers that may or may not have GPUs in a synchronous manner. It is similar to `MirroredStrategy` but differs in the fact that it implements an **All-Reduce** execution mode to be able to make the nodes work together. The code is shown here:

```
def train_custom_strategy():
    import tensorflow as tf
```

```python
   strategy = tf.distribute.experimental.
MultiWorkerMirroredStrategy(
      tf.distribute.experimental.CollectiveCommunication.NCCL)
  with strategy.scope():
    import uuid
    BUFFER_SIZE = 10000
    BATCH_SIZE = 32
    def make_datasets():
      (mnist_images, mnist_labels), _ = \
        tf.keras.datasets.mnist.load_data(path=str(uuid.
uuid4())+'mnist.npz')

      dataset = tf.data.Dataset.from_tensor_slices((
          tf.cast(mnist_images[..., tf.newaxis] / 255.0,
                  tf.float32),
          tf.cast(mnist_labels, tf.int64))
      )
      dataset = dataset.repeat().shuffle(BUFFER_SIZE).
batch(BATCH_SIZE)
      return dataset


    def get_model():
      model = tf.keras.Sequential([
        tf.keras.layers.Conv2D(32, 3, activation='relu',
                               input_shape=(28, 28, 1)),
        tf.keras.layers.MaxPooling2D(),
        tf.keras.layers.Flatten(),
        tf.keras.layers.Dense(138, activation='relu'),
        tf.keras.layers.Dense(10, activation='softmax'),
      ])
      model.compile(
        loss=tf.keras.losses.sparse_categorical_crossentropy,
        optimizer=tf.keras.optimizers.SGD(learning_rate=0.001),
        metrics=['accuracy'],
      )
      return model
```

```
train_datasets = make_datasets()
worker_model = get_model()

options = tf.data.Options()
options.experimental_distribute.auto_shard_policy = \
    tf.data.experimental.AutoShardPolicy.DATA
train_datasets = train_datasets.with_options(options)
worker_model.fit(x=train_datasets, epochs=3,
                 steps_per_epoch=5)
```

Now, we can distribute the training of the model by running
`MirroredStrategyRunner` with the `local_mode` and `use_custom_strategy`
parameters set to `True` for local execution of the specified training strategy, as follows:

```
runner = MirroredStrategyRunner(num_slots=1,
                                use_custom_strategy=True,
                                local_mode=True,
                                use_gpu=USE_GPU)
runner.run(train_custom_strategy)
```

We can run this in local mode to verify that the script works properly before scaling the
training. We can check this by verifying that `CollectiveCommunication.NCCL` is
printed in the logs.

As we can see, the Spark TensorFlow Distributor is an alternative to `HorovodRunner`
for managing distributed training of TensorFlow deep learning models. We can use this
library to train the models both on a single node and in a cluster in a very simple way,
being able to harness the power of a GPU or CPU cluster for distributed training of deep
learning models.

Summary

We have learned in this chapter about how we can improve the performance of our
training pipelines for deep learning algorithms, using distributed learning with Horovod
and the native TensorFlow for Spark in Azure Databricks. We have discussed the core
algorithms that drive the capability of being able to effectively distribute key operations
such as gradient descent and model weights update, how this is implemented in the
`horovod` library, included with Azure Databricks Runtime for Machine Learning, and
how we can use the native support now available for Spark in the TensorFlow framework
for distributed training of deep learning models.

This chapter concludes this book. Hopefully, it enabled you to learn in an easier way the incredible number of features available in Azure Databricks for data engineering and data science. As mentioned before, most of the code examples are modifications of the official libraries or are taken from the Azure Databricks documentation in order to provide well-documented examples, so kudos for the hard work to all of the developers that help build these libraries and tools.

Packt.com

Subscribe to our online digital library for full access to over 7,000 books and videos, as well as industry leading tools to help you plan your personal development and advance your career. For more information, please visit our website.

Why subscribe?

- Spend less time learning and more time coding with practical eBooks and Videos from over 4,000 industry professionals
- Improve your learning with Skill Plans built especially for you
- Get a free eBook or video every month
- Fully searchable for easy access to vital information
- Copy and paste, print, and bookmark content

Did you know that Packt offers eBook versions of every book published, with PDF and ePub files available? You can upgrade to the eBook version at packt.com and as a print book customer, you are entitled to a discount on the eBook copy. Get in touch with us at customercare@packtpub.com for more details.

At www.packt.com, you can also read a collection of free technical articles, sign up for a range of free newsletters, and receive exclusive discounts and offers on Packt books and eBooks.

Other Books You May Enjoy

If you enjoyed this book, you may be interested in these other books by Packt:

Azure Data Factory Cookbook

Dmitry Anoshin, Dmitry Foshin, Roman Storchak, Xenia Ireton

ISBN: 978-1-80056-529-6

- Create an orchestration and transformation job in ADF
- Develop, execute, and monitor data flows using Azure Synapse
- Create big data pipelines using Azure Data Lake and ADF
- Build a machine learning app with Apache Spark and ADF
- Migrate on-premises SSIS jobs to ADF
- Integrate ADF with commonly used Azure services such as Azure ML, Azure Logic Apps, and Azure Functions
- Run big data compute jobs within HDInsight and Azure Databricks
- Copy data from AWS S3 and Google Cloud Storage to Azure Storage using ADF's built-in connectors

Azure Data Engineering Cookbook

Ahmad Osama

ISBN: 978-1-80020-655-7

- Use Azure Blob storage for storing large amounts of unstructured data
- Perform CRUD operations on the Cosmos Table API
- Implement elastic pools and business continuity with Azure SQL Database
- Ingest and analyze data using Azure Synapse Analytics
- Develop Data Factory data flows to extract data from multiple sources
- Manage, maintain, and secure Azure Data Factory pipelines
- Process streaming data using Azure Stream Analytics and Data Explorer

Packt is searching for authors like you

If you're interested in becoming an author for Packt, please visit `authors.packtpub.com` and apply today. We have worked with thousands of developers and tech professionals, just like you, to help them share their insight with the global tech community. You can make a general application, apply for a specific hot topic that we are recruiting an author for, or submit your own idea.

Leave a review - let other readers know what you think

Please share your thoughts on this book with others by leaving a review on the site that you bought it from. If you purchased the book from Amazon, please leave us an honest review on this book's Amazon page. This is vital so that other potential readers can see and use your unbiased opinion to make purchasing decisions, we can understand what our customers think about our products, and our authors can see your feedback on the title that they have worked with Packt to create. It will only take a few minutes of your time, but is valuable to other potential customers, our authors, and Packt. Thank you!

Index

www.ingramcontent.com/pod-product-compliance
Lightning Source LLC
LaVergne TN
LVHW081329050326
832903LV00024B/1086